Creative Perspectives on Sustainable Nature–Culture Relationships

Developing a new understanding of the cultural relationships of humans with nature through heritage and tradition, this book examines ways of transforming these relationships sustainably and assesses the impact on policy-related actions for ecologically embedded futures, drawing particularly on case studies from the European north.

This book opens up fresh perspectives, drawing on current debates in multispecies studies and bringing together approaches from environmental activism to community art practice and critical heritage studies. Focusing on ecological relationships underpinning the making of meaningful places, it combines theoretical exploration with case studies that raise themes, issues, and eco-cultural practices that have relevance across the world. Contributors evaluate intangible cultural heritage and traditional ecological knowledge for their potential to augment sustainable development, thus encouraging the rethinking of sustainability as a human-ecological concept indicating how new, locally rooted relationships with nature may create policy-oriented pathways towards cultural sustainability transformation.

Creative Perspectives on Sustainable Nature–Culture Relationships provides new insights to both professionals and students in the fields of sustainability studies, involving human ecology, geography, landscape studies, cultural anthropology, cultural heritage studies, art, archaeology, performance, literature, linguistics, futures research, philosophy, and beyond.

Ullrich Kockel is Professor of Creative Ethnology at the Institute for Northern Studies, University of the Highlands and Islands, Scotland, and a Visiting Professor at Vytautas Magnus University Kaunas, Lithuania, and the Latvian Academy of Culture. A member of the Royal Irish Academy, the Steering Group of Learning for Sustainability Scotland, a United Nations University-recognised Regional Centre of Expertise, and the Scottish Centre for Geopoetics, he works on issues of place, cosmology, cultural ecology, and sustainability.

Riina Haanpää is university lecturer and Associate Professor in Cultural Heritage Studies at the University of Turku, Finland. Her research focuses on oral history and memory as well as local and traditional ecological knowledge. Haanpää has worked on several local oral history projects in the Satakunta region and developed community-based oral history research.

Mairéad Nic Craith is Professor of Public Folklore at the Institute for Northern Studies, University of the Highlands and Islands in Scotland. A member of the Royal Irish Academy, her research seeks to integrate critical heritage studies, cultural history, literature, and folklore into a creative ethnology. Author of seven monographs and sole or joint editor of eleven collections, her most recent monograph is *The Vanishing World of The Islandman: Narrative and Nostalgia* (2020). Her current research focuses on environmental aspects of Irish-Gaelic mythology and folklore.

Katriina Siivonen is Associate Professor of Cultural Heritage Studies and senior university lecturer in Futures Studies at the University of Turku, Finland. She combines futures research, cultural heritage research, and sustainability science into a transdisciplinary entity in her work. Through more than 25 years of research, she has developed the study of cultural sustainability transformation and heritage futures into a thematic research area integrated into society. She is a member of the Expert Panel for Sustainable Development in Finland, and has been chairing the Advisory Board of the implementation of the UNESCO Convention for the Safeguarding of the ICH in Finland.

Routledge Studies in Culture and Sustainable Development

Series Editors: Nancy Duxbury, *Centre for Social Studies at the University of Coimbra, Portugal and* **Julius Heinicke,** *Institute of Cultural Policy, University of Hildesheim*

Culture as an aspect of sustainability is a relatively new phenomenon but is beginning to attract attention among scholars and policy makers. This series opens up a forum for debate about the role of culture in sustainable development, treating culture and sustainability as a meta-narrative that will bring together diverse disciplines. Key questions explored in this series will include: how should culture be applied in sustainability policies; what should be sustained in culture; what should culture sustain; and what is the relationship of culture to other dimensions of sustainability?

Books in the series will have a variety of geographical foci and reflect different disciplinary approaches (for example, geography, sociology, sustainability science, environmental and political sciences, anthropology, history, archaeology and planning). The series will be addressed in particular to postgraduate students and researchers from a wide cross-section of disciplines.

Performing Sustainability in West Africa
Cultural Practices and Policies for Sustainable Development
Edited by Meike Lettau, Christopher Yusufu Mtaku and Eric Debrah Otchere

Cultivating Islamic Cultural Capitals
Educational Empowerment under ICESCO
Mustafa Osman I. Elamin and Abdelaziz Khatir Osman Jadeed

Creative Perspectives on Sustainable Nature–Culture Relationships
Beyond the Anthropocene
Edited by Ullrich Kockel, Riina Haanpää, Mairéad Nic Craith, and Katriina Siivonen

For more information about this series, please visit: www.routledge.com/Routledge-Studies-in-Culture-and-Sustainable-Development/book-series/RSCSD

Creative Perspectives on Sustainable Nature–Culture Relationships

Beyond the Anthropocene

Edited by
Ullrich Kockel, Riina Haanpää,
Mairéad Nic Craith, and Katriina Siivonen

Routledge
Taylor & Francis Group
LONDON AND NEW YORK

earthscan
from Routledge

First published 2027
by Routledge
4 Park Square, Milton Park, Abingdon, Oxon OX14 4RN

and by Routledge
605 Third Avenue, New York, NY 10158

Routledge is an imprint of the Taylor & Francis Group, an informa business

British Library Cataloguing-in-Publication Data
A catalogue record for this book is available from the British Library

ISBN: 9781041215875 (hbk)
ISBN: 9781041215899 (pbk)
ISBN: 9781003727972 (ebk)

DOI: 10.4324/9781003727972

Typeset in Sabon
by codeMantra

Contents

Figures

Contributors

Iain Biggs, RWA, is an independent writer, artist, consultant, and former director of the PLaCE Research Centre, University of the West of England, Bristol. He is co-coordinator of PLaCE International, co-author of *Creative Engagements with Ecologies of Place: Geopoetics, Deep Mapping and Slow Residencies* (2020), and associate of the Temporal School of Experimental Geography.

Patrick Dillon is Emeritus Professor in the Faculty of Arts, Humanities, and Social Sciences at the University of Exeter, England, and a cultural ecologist. He holds elected positions at the Universities of Eastern Finland and Helsinki. His book *Cultural Ecologies of the Land: A Restless Dynamic of People and Place* was published in 2025.

Maarit Grahn is a museum assistant in the city of Ulvila, Finland. She received a doctorate in Cultural Heritage Studies from the University of Turku and has worked on several local history and industrial heritage projects in the Satakunta region. In addition, her research focuses on cultural heritage entrepreneurship and history management.

Riina Haanpää is university lecturer and Associate Professor of Cultural Heritage Studies at the University of Turku, Finland. Her research focuses on oral history and memory as well as local and traditional ecological knowledge. Haanpää's work covers several local oral history projects in the Satakunta region and community-based oral history research.

Ullrich Kockel is Professor of Creative Ethnology at the Institute for Northern Studies, University of the Highlands and Islands, Scotland. A member of the Royal Irish Academy and the Steering Group of Learning for Sustainability Scotland, a United Nations University-recognised Regional Centre of Expertise, he has published extensively on issues of place, cosmology, cultural ecology, and sustainability.

Pauliina Latvala-Harvilahti is senior research fellow in European Ethnology and Associate Professor of Cultural Heritage Studies at the University of Turku, Finland, and project manager at the Finnish Museums Association.

Her research spans from oral history to heritage policy, from artivism to future agency, and cultural sustainability transformation, including several authored and edited publications.

Kirsi Laurén is senior lecturer in Cultural Heritage at the University of Oulu, Finland, and Associate Professor of Environmental Humanities at the University of Eastern Finland. Her research interests include human–nature relationship, and the study of narratives, cultural heritage, and oral history. She has authored several publications and edited thematic issues for scientific journals.

Mary Modeen is Professor of Interdisciplinary Art Practice at Duncan of Jordanstone College of Art and Design, University of Dundee, Scotland. She combines creative work with academic study and writing across several disciplines. Her publications include *Creative Engagements with Ecologies of Place: Geopoetics, Deep Mapping and Slow Residencies* (2020).

Mairéad Nic Craith is Professor of Public Folklore at the Institute for Northern Studies, University of the Highlands and Islands, Scotland, and a member of the Royal Irish Academy. Her monograph *The Vanishing World of The Islandman* (2020) critically integrates heritage, history, folklore, and literature. She is currently researching environmental aspects of Gaelic folklore and mythology.

Laura Puolamäki is senior researcher in Landscape Studies at the University of Turku, Finland. Interested in local environmental knowledge, sustainability, and cultural landscape, her work covers research and development projects related to landscape, tourism, environment, and cultural heritage. She is a member of the Baltic University Programme's research network.

Eeva Raike is university teacher in Landscape Studies at the University of Turku, Finland. She is doctoral researcher in the Degree Programme of Digital Culture, Landscape Studies, and Cultural Heritage. She is an archaeologist and interested in old maps, landscape archaeology, and history. Currently, her focus is on historical rock carvings.

Katriina Siivonen is Associate Professor of Cultural Heritage Studies and senior university lecturer in futures studies at the University of Turku, Finland. She combines futures research, cultural heritage research, and sustainability science in her transdisciplinary work. She is a member of the Expert Panel for Sustainable Development in Finland.

Priscilla Trenchard is a freelance multidisciplinary artist specialising in print and collage. Most of her work is abstract. She has been artist in residence on archaeological excavations producing work in and with the environment. She is a member of the creative practice-led research network LAND2.

Preface

In the aftermath of a workshop on 'Between Habitus and Habitat: Steps towards EcoEthnology' at the 2008 Congress of the *Société Internationale d'Ethnologie et de Folklore* (SIEF), which took place at the University of Ulster's Magee campus, a SIEF Working Group on Place Wisdom was initiated by researchers working at the Academy for Irish Cultural Heritages (University of Ulster), the Ethnological Research Unit (University of the West of England, Bristol), and the Centre for Human Ecology (then attached to the University of Strathclyde). Following approval by SIEF's Executive Board, the group was officially launched at the University of Ulster on 1 May 2009 with a symposium on 'community, ecology and the human spirit', organised by Ullrich Kockel, at the time Professor of Ethnology and Folklife at Ulster and founding-convenor of the group; selected papers from this symposium were published as a special issue of the *Anthropological Journal of European Cultures* (2010). Workshops, panels, and roundtables on Place Wisdom themes have been organised at every SIEF Congress since then, with smaller events, during inter-congress periods, in Bristol (2010), Mannheim (2014), Wrocław (2015), and Perth/Scotland (2022).

At the 2015 SIEF Congress in Zagreb, Ullrich Kockel handed on the baton to Stella Butter (Germany) and Katriina Siivonen (Finland) as joint convenors of the group. They were succeeded, in due course, by Cáit O'Neill McCullagh (Scotland), Pauliina Latvala-Harvilahti (Finland), Victoria Walters (England), and Kirsty Kay (England). At the time of writing, the joint convenors, elected at the 2025 SIEF Congress in Aberdeen, are Anne Dippel (Germany) and Sarah Squire (Scotland).

The group's aim is to promote research in relation to understandings of place. Involving the creation and gathering of ethnographically inflected multimedia resources, a particular focus of the group's research is on ecological beliefs and practices and traditions that contest conventional divisions between culture and nature. Balancing on the edges between anthropology, art, futures studies, geography, human ecology, literature, performance studies, and philosophy, the group's work has been concerned with issues such as the spiritual foundations of community development or the meaning of indigeneity in the post-industrial world. Using ethnographic and closely related local case studies to challenge received notions of 'tradition', the members of the group have examined 'tradition' as a process of intergenerational, inter-regional, and intercultural transfer of knowledge, skills, and values, looking at the creative potential of living traditions in their socio-ecological contexts, and at the role of the ethnologist in the tradition process. Located at the interface of ethnology and other arts and humanities fields, the research has from the start involved a significant element of practice in various forms, including audio-visual, fine art, and performance.

This book was inspired by the panel on 'New rules for the engagement with nature: human ecology and emerging heritage futures', hosted by the group at the 2021 SIEF Congress in Helsinki. The book expands on themes raised in that panel and in subsequent seminars and other engagements between group members and their colleagues and collaborators.

We are grateful to all who contributed to this debate, which included the constructive input of anonymous peer reviewers.

As with all such projects, thanks are also due to many for supporting our endeavour to extend the debate beyond the seminar room: Routledge editors Grace Harrison and Morwenna Scott guided us through the commissioning process; series editors Nancy Duxbury and Julius Heinicke welcomed the book in their series; the final touches to the text were assisted by Rohita Divyanshu and Hannah Studd at Routledge, and Romaine Heldt at codeMantra oversaw production. Without their help our book would not have been possible.

We dedicate this book with much love to Cáit O'Neill McCullagh, distinguished poet and essayist, an experienced educator in community and higher education settings in Scotland. Cáit's research and her enthusiasm for the Place Wisdom working group have been an inspiration to all of us. Her pamphlet of poems, *The Songs I Sing Are Sisters*, co-authored with Sinéad McClure, was co-winner of Dreich's 'Classic Chapbooks' 2022 and won a Saboteur Award in 2023. *The Bone Folder*, her first solo, full-length collection of poems, was published in 2024 by Drunk Muse Press and has been longlisted for the Saltire Book Award in 2025. Cáit has performed her poetry at numerous national and international poetry gatherings. Her poetry has been nominated for a Pushcart Poetry Prize and earned her recognition with diverse organisations, including as Inverness Skatepark's inaugural 'Poet in Motion', the Scots Language Centre's 'Makar o the Muinth', and most recently poet in residence at the Institute for Northern Studies, University of the Highlands and Islands. Cáit is surviving a diagnosis of cancer, identified in the spring of 2022.

1 Creative Perspectives on Sustainable Nature–Culture Relationships

Setting the Context

Ullrich Kockel, Mairéad Nic Craith, Katriina Siivonen, and Riina Haanpää

Philosophers have long argued that language, as it shapes our projections, both facilitates and confines our understanding of the world. As researchers grapple with complex phenomena, they regularly resort to neologisms that they hope will further their – and, by extension, everyone's – understanding of the world. In 2000, the Nobel laureate Paul Crutzen (Crutzen and Stoermer 2000), recognising that human influences were significantly altering planetary conditions, postulated such a term to describe the current geological epoch – the Anthropocene (from Old Greek: *ἄνθρωπος*, ánthropos, 'human being', and *καινός*, kainós, 'new'). This label captures the perception that unless humans revise their relationships with nature, damage to the Earth will be irredeemable. The present volume considers the cultural dimensions to such revised relationships, including ways of transcending the nature–culture binary that the Anthropocene concept implies.

Towards a Radical Human Ecology

The idea of the Anthropocene was not entirely new. Similar concepts had been circulating in environmental discourse at least since it emerged beyond its romantic origins (see, e.g., Wulf 2016) during the 1960s. For much of the first quarter of the new century, the Anthropocene became the master narrative of the expanding environmental humanities. It encompassed several component parts and contributory factors, from Empire and Globalisation, to various sub-cenes, such as the Capitalocene (see Moore 2016). As this metaphysical universe of discourse expanded, critical voices were raised at its periphery, pointing out that not only was the Anthropocene a highly anthropocentric concept – it also implied that the problems it tried to capture were best solved by hi-tech human interventions in natural processes (Jenkins 2023). In short, it maintained the juxtaposition of (civilised forms of) humanity and its natural environment – the much maligned 'culture vs nature' binary. The challenge to overcome this binary encouraged 'lexical struggles to find vocabulary for "multispecies" or "more-than-human" relations' (Jenkins 2023, 622), producing various approaches inspired by thinkers like Donna Haraway (2007) with her influential concept of 'natureculture'.

DOI: 10.4324/9781003727972-1

The proliferation of terms has helped open up new perspectives on the old binary, which remains unresolved, however, because of a core issue noted by Milstein (2024) – language shapes our world. The stories we tell ourselves and one another form our actuality; but can they really put us into the position of a bee, a birch, a brook, or a boulder? If not, are environmental humanities just engaging in anthropomorphism when they advocate seeking such empathy? And would that really be such a bad thing to do?

In their introduction to the *Routledge Handbook of Ecocultural Identity*, Tema Milstein and José Castro-Sotomayor (2020, xvii) have pointed out that for an ever-growing majority of humans, 'it seems as if our ecological selves have become steadily less accessible', leading to 'a lack of earthly self-awareness in an increasingly human-centered world'. This increasing alienation becomes obvious in 'the invisibility and deniability we assign to our environmental interlinkages, impacts, and interdependencies'. Critiquing the concept of the Anthropocene, the German philosopher Jürgen Manemann (2014) has expressed this situation in terms of the contrast between *Menschenflucht* – escaping from being (an ecologically connected) human – and *Menschwerdung* – becoming fully human. The former morphs into 'Hominisation', the expansion of human dominance over both 'internal' and 'external' nature. Manemann called for this dynamic to be overcome through 'humanisation', human self-actualisation based on humility and compassion. This concept resonates with David Abram's idea of the 'Humilocene' (Abram et al. 2020).

Manemann's argument – that the path of societal development is determined between these two polar opposites – undermines the continued focus on data and scenarios dominating a science- and technology-fixated discourse; it thus critiques the implicit anthropology of the Anthropocene concept. The ethical imperative that this concept entails is derived from an impulse for societal transformation that comes with the acute observation of environmental change that is attributable to human impact on the ecosystem. At an early stage, the discourse on the Anthropocene moved beyond description and classification towards a normative dimension, turning into a discourse of responsibility for the Earth (e.g., Rockström et al. 2009). This led to a revival of an image from cultural history: the gardener as steward of the Earth who creates order and progress, not least by cultivating new varieties and species. At this point, Manemann's critical analysis casts the idea of the Anthropocene as naïve progress euphoria linked to an ideology of achievability. There is, however, more to the environmental humanities than the Anthropocene discourse. And this takes us closer to the kind of radical human ecology Manemann postulates – but not without opening another discursive minefield or two.

Whereas critical modifications of the Anthropocene concept, such as the Capitalocene, point the searchlight at specific cultural expressions associated with the philosophical imaginary of 'economic man' – including commercialisation, industrialisation, and progress-driven abuse of resources – and date its onset accordingly (Graeber and Wengrow 2022), Indigenous[1] scholars (e.g., Davis and Todd 2017; Whyte 2017) have suggested to date the environmental crisis back to the colonisation of the Americas. Their critique highlights other aspects of current mainstream understandings of the Anthropocene concept – such as the ontological binary of nature and society, the individualist and anthropocentric focus, and the framing of the environmental crises in primarily scientific terms – as rooted in colonialism and reinforcing systems of post-colonial domination. This leads to calls for indigenisation and decolonisation of the Anthropocene concept. Moreover, the significance of spiritual connection to the land has been highlighted as crucial for any ecological movement (e.g., Wildcat 2009), and this emphasis infuses proposals for a radical human ecology (e.g., Williams et al. 2012) that have emerged at the same time as issues of place and kinship, and the exploration of new relationships between and new ways of thinking about them, have drawn attention to the concept of 'indigeneity' and its application in contexts beyond those conventionally associated with it. The concept has been instrumentalised by right-wing populists, and commercially exploited by the heritage and tourism industry in particular, which has caused friction (Kockel 2012, 2024).

High-profile cases of individuals from the mainstream culture misappropriating Indigenous identity to authorise their claims to Indigenous dignity and the exploitation of traditional knowledge have aggravated the broader global question of how those described as 'no longer Indigenous to place' may communicate – in the widest sense of that term – with Indigenous peoples. In academic contexts, this has generated some debate over whether non-Indigenous academics have any standing when it comes to researching, writing about, and teaching on Indigenous issues and knowledge systems. The contributors to a recent collection of essays, mainly with an Aboriginal and Indigenous law background, consider non-Indigenous scholars as having a duty to engage, not so much in terms of 'what is their *place*, but rather what are their *responsibilities*' (Borrows and McNeil 2022, 10; orig. emph.), and to do so in a respectful manner, speaking with, rather than at, for, or merely to one another (see also Williams et al. 2024). In a European context, these thoughtful essays have relevance for the growing debate on 'indigeneity' in situations shaped by internal colonialism, which may well expand the scope of the concept beyond its compass established by international conventions (Kockel 2024).

The people of 'internal colonies' of Europe share with Indigenous peoples the experience of being stereotyped as exotic 'tribes' that are socially marginal, geographically peripheral, and culturally primitive. With reference to the Sámi, Löfgren (1996, 158) noted the historical dispute that arose in Sweden over 'who had a right to the Lapps': as a part of Swedish national heritage, their ethnographic representation would have fallen within the remit of the Nordic Museum; as an 'exotic tribe', they would have 'belonged with the other primitives of the zoology collections' in the Natural History Museum.

Many years after that dispute, the Sámi artist and activist Nils-Aslak Valkeapää (1983, 3) noted caustically that 'an average Sami family consists of parents, three children and an anthropologist' – an observation that could equally have been made in the West of Ireland, which was considered 'exotic' enough to be of interest for anthropology long before that discipline's 'home-coming' to 'civilised' locations. Both the museum and classical anthropological fieldwork are part of the Western, post-Enlightenment approach to knowledge that is deeply implicated in colonialism and the Capitalocene (Pels 2022), yet while the Sámi are recognised as an Indigenous people, the Irish-speakers in the West of Ireland are not. Both would, arguably, benefit from 'full acceptance of other, alternative epistemologies as being equal to Western systems of knowledge within the academia' as part of what Kuokkanen (2000, 416) cautiously refers to as an 'Indigenous Paradigm'.

Alternative epistemologies – and the cosmologies they draw on and build – have long existed in the cultural context that we tend to refer to as 'Western', but they have been marginalised by the hegemonic triad of science, capital, and empire. Confronted early on with ecologically devastating impacts of the hegemonic triad, the Romantics reacted against the spirit of global conquest spurred by the Enlightenment, or perhaps one should better say: against its being misappropriated by political-economic interests. Alexander von Humboldt was one of the first to turn away from the objectivising analytical method of looking at facts in isolation, declaring that everything was 'interaction and reciprocal' (quoted in Wulf 2016, 59). He considered 'imagination [...] as necessary as rational thought in order to understand the natural world' (Wulf 2016, 36).

To English Romantics and American Transcendentalists, subjective ways of knowing – through art, poetry or intuition – were a means of recovering the unity of the inner world of humanity with the outer world of nature – not pitched against each other, but each an integral part of the other. While the objective–subjective binary aligns this worldview with the classical philosophical dualism of realism vs idealism – the former being rated more highly by the hegemonic triad – the cosmology this epistemology engenders is similar in many ways to the 'Indigenous paradigm' championed by Kuokkanen (2000). The growing interest in traditional ecological knowledge (TEK) in and beyond the lifeworld of peoples recognised as Indigenous is therefore not surprising. Exploring these similarities with reference to other indigenous cultural contexts can help the cultivation of new relationships with nature as long as it is carried out in a spirit of responsibility, informed by acute awareness of the complicity with the colonial project many dispossessed indigenous peoples from the margins of Europe have been forced into (see, e.g., Ignatiev 1995; Englert 2022).

Traditional knowledge and practices lie at the heart of a community's culture and identity but are under serious threat from globalisation. While indigenous cultures around the globe have experienced the loss of their TEK, Indigenous peoples have been particularly affected by, for example, pharmaceutical corporations interested in the medicinal uses of local plant species in the Amazon. Clearing of land for development may result in the disappearance of a sacred grove or the need to find an alternative source of wood for building. Climate change due to deforestation and the spread of deserts inevitably threatens many already endangered species. This results in the decline of, for example, traditional crafts and herbal medicine as raw materials and plant species disappear. The focus of this book is on voluntary, conscious transformation of cultural practices that may provide the foundations for new natureculture relationships, a radical human-ecological transformation regarded as necessary to tackle the ecological crisis. While 'tackling the crisis' is unlikely to stop or even reverse climate change at least in the short run – and in the long run, as economist John Maynard Keynes reputedly once observed: we will all be dead – these transformed cultural practices, if they evolve into living traditions, may well provide the grounding for sustainable indigenous responses to the crisis.

Outline of This Book

The contributions in this collection focus on traditions associated with features of nature, such as forests, mires and peatlands, mountains, lakes, rivers, and maritime environments, extending beyond history and memory, including towards imagining and assembling futures. These traditions are aspects of cultural heritage, including Heritage Futures (Kockel 2019), that can add understanding to our relationships with nature, including our observations of and responses to environmental changes. Human–nature relationships are affected by changes in society. However, these relationships can also be co-created to empower people to change societal structures and practises. Because of the ecological crisis, we need to think more seriously than ever about what kind of interactions between humans and nature may be sustainable. In this cultural-ecological context, natural areas have become physical and symbolic sites not only for travel, recreation, or documentation, but for active interventions conveying political messages and creative dialogue examining possibilities for sustainable cultural transformation. The book explores several themes arising in this context that infuse the various chapters.

What Constitutes Liveable Ecological Futures?

Milstein (2024) reminded us that, with the greatest empathy imaginable, we are not able to leave our ecological vantage point as human beings. While our engagement with the world therefore remains inevitably anthropomorphic, it does not necessarily make our vantage point anthropocentric in any other than a strictly geometrical sense. Like all other members of the multispecies cosmos, we are relational beings, not pitched against or positioned hierarchically above, but living within nature, concerned with respect for and the sustainability of relations that support this livelihood. Such a radical human ecology is facilitated by what Ivan Illich more than half a century ago described as 'deschooling society' (Illich 1971), an alternative pedagogy that will give us 'tools for conviviality' (Illich 1973), a transformed approach to technology. While Illich's work coincided with the early stirrings of the ecology movement and concentrated narrowly on societal issues, it significantly inspired more contemporary movements, such as the *Dark Mountain Project* (Kingsnorth and Hine 2009), which regard technological solutions to climate change as not possible, suggesting instead that notions of 'progress' ought to be re-evaluated. Underpinning the quest for liveable ecological futures is a key concern: How might we cultivate new relationships with nature? The answer(s) will in large part depend on how we define 'futures', and what cultural transformations leading us towards these are feasible.

One of this book's editors, Katriina Siivonen, has argued that culture itself is an interactive part of nature (e.g., Siivonen 2018, 2022). Siivonen and Kockel's essay in this volume aims to develop a conceptual model of Heritage Futures that combines (1) transformative power of culture defined as a dynamic, interactive process that is a part of nature; (2) anticipatory understanding; and (3) cultural heritage as a tool to engage people in an inspiring, affectual, cognitive and practical way (see also Siivonen 2022). The model of Heritage Futures is an intentional, anticipatory, cultural tool to co-create alternative images for sustainable futures, and accordingly skills and actions in culture and society.

Cultural sustainability transformation meets complex systems at all leverage points (Siivonen 2022; Meadows 1999). This brings to the fore a need to balance different ethical standpoints of several root disciplines that meet in an interdisciplinary policy context where joint, urgent problems are researched. A transdisciplinary approach, combining different cultural and societal actors and activities with interdisciplinary spheres, adds additional ethical demands. Anthropological and ethnological professional ethics claim strongly that researchers need to respect individuals and communities involved in the research processes. Researchers should not intervene in the cultural phenomena of the research field, and should support the cultural rights of people belonging to it. However, a policy of cultural sustainability transformation requires intentional, goal-driven cultural change in human-nature relationships and value-based, cultural habits and practices of overconsumption. Thus, in transdisciplinary research, the ethical guidelines of ethnology and anthropology may need to be revisited, to deal with the applied dimension. Poli (2011) argues that everyone, including the policy researcher, has a responsibility to respect the future of both nature and human life. The responsibility of the professional Futures researchers lies in the support given to people to 'become aware of the values that guide their actions through the consequences that may arise from them' (Poli 2011, 409). In transdisciplinary research, such as in co-creation of cultural sustainability transformation, there is a need to find a balance between different professional ethical standpoints (Siivonen 2022).

The core of cultural sustainability can be seen as the right of people to take part in, and have an impact, either on their own culture – understood as framework (see, e.g., Geertz 1957) – or on cultural change in their own cultural environment – the metaphor of culture as river (see Hannerz 1992). However, this does not guarantee the direction of cultural change towards sustainability of any dimension – ecological, economic, social, or cultural. The possibility to take part and have an impact can, however, be seen as a condition of cultural transformation towards sustainability (Siivonen 2008, 2017, 2022).

In order to help us better understand the human–nature interface, and to form practical solutions towards a more sustainable world, Heritage Futures needs to include the experience of both cultivated and un-cultivated nature. Siivonen (2022) argues that Heritage Futures require disruptive reflection and co-creation of our daily cultural habits, and care for their sustainable impacts on nature and human beings with respect for planetary boundaries. Beyond the external challenges to the social and natural environment, many underprivileged or marginalised communities are inclined to adopt a way of life or an economic development model far from their own traditions and customs. Ideas of protecting the natural environment, while inadvertently perpetuating the culture–nature binary, are often linked to safeguarding a community's cosmology, as well as other examples of its intangible cultural heritage (ICH).

TEK as ICH

Worldviews, knowledge, skills, actions, habits, and practices, including understanding of the place of humans in nature, belong to intangible culture. In social, economic, and political contexts, these are identified as ICH. Although the definition of ICH remains ambiguous and indeed contested (Noyes 2015; see also West 2012), it has been customary to follow usage established by UNESCO (see, e.g., Kockel and McFadyen 2019; Nic Craith et al. 2019). In the Convention for the Safeguarding of the ICH (UNESCO 2003), ICH has been defined as 'the practices, representations, expressions, knowledge, skills – as well as the instruments, objects, artefacts and cultural spaces associated therewith – that communities, groups and, in some cases, individuals recognize as part of their cultural heritage'. Adopting this definition, UNESCO drew on discussions that had previously taken place in folklore studies. The Convention has directed attention to different forms of ICH and sub-divided them into five broad domains, illustrated in Figure 1.1.

Inevitably, this draws on traditional ecological wisdom. As examples of this, UNESCO cites 'knowledge about local fauna and flora, traditional healing systems, rituals, beliefs, initiatory rites, cosmologies, shamanism, possession rites, social organisations, festivals, languages and visual arts'. TEK is a flexible term, and there is no single agreed universal definition (Berkes 1993). It includes the knowledge of farmers, fishermen, craft workers, and

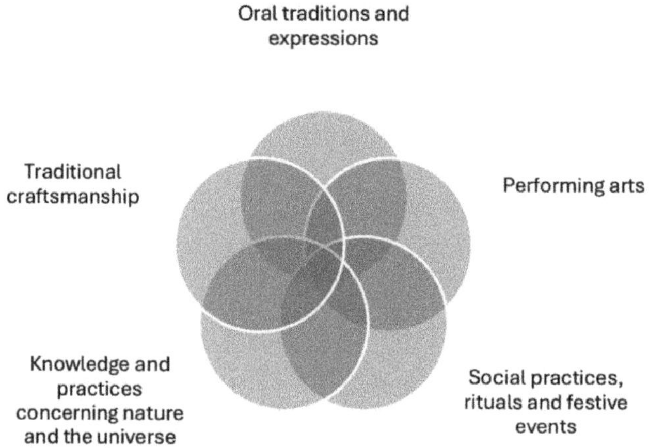

Figure 1.1 Subdivisions of intangible cultural heritage.

Source: Figure by Nic Craith (Nic Craith and McDermott 2025, 64).

tacit knowledge of the environment. Drawing on Berkes (1993), Mazzocchi (2006, 463) defines it as 'a cumulative body of knowledge, practices and representations that describes the relationships of living beings with one another and with their physical environment, which evolved by adaptive processes and has been handed down through generations by cultural transmission'.

Many of the examples of TEK and ICH in this volume fall primarily into the ICH domain entitled 'knowledge and practices concerning nature and the universe'. This domain includes 'knowledge, knowhow, skills, practices and representations developed by communities by interacting with the natural environment' (UNESCO n.d.). The definition is quite broad-ranging and refers to elements of place-based knowledge and skills that are expressed in different ways, such as language, memories, and spirituality. These often emerge from specific places and worldviews and have been strongly influenced by values and beliefs.

Examples of cultural practices entered on the representative list in this domain include Bear festivities in the Pyrenees, and *Chogān*, a horse-riding game in Iran, accompanied by music and storytelling. Many seasonal festivals feature on the list, such as traditional spring festive rites of the horse breeders in Kazakhstan or the Spring celebration *Hıdrellez* in Turkey and North Macedonia. Examples of water management are also listed, including the traditional system of Corongo's water judges (Peru) and Al Aflaj, a traditional irrigation network system in the United Arab Emirates, including oral traditions, knowledge and skills of construction, maintenance, and equitable water distribution. In Finland, sauna culture has recently been included in the list, and human connection to nature through sauna bathing is described as a part of it. However, the heating of saunas as a part of everyday life involves a remarkable proportion of the energy consumption of a household in a country with 3.3 million saunas for 5.5 million inhabitants. Alongside its positive cultural values, sauna bathing can be seen as one part of over-consuming natural resources (energy) and causing environmental problems.

Unlike typical Western reliance on perceived objectivity and truths, TEK varies from place to place and reflects diverse worldviews in different locations. Much of this wisdom has been transmitted orally, and in habits and practices as tacit knowledge over many generations (Mazzocchi 2006). Many of these habits and practices have emerged from traditional systems of beliefs that have helped, for instance, Indigenous peoples to interpret and nurture the physical environment (Iaccarino 2003). This knowledge is derived from long-term empirical observations in local contexts. It is a form of engagement that does not drain resources but considers the needs of future generations.

Such engagement with nature is not confined to those legally designated Indigenous in a European context (see Kockel 2024). Mairéad Nic Craith's contribution to this volume explores the indigenous Irish-Gaelic engagement with nature in a pre-colonial context and asks whether that pre-colonial relationship can be retrieved or revived with a view towards more ecological natureculture futures. The value of traditional knowledge for biodiversity is increasingly recognised (Gadgil et al. 1993). Its significance is not simply a mechanism to slow down the loss of biodiversity in an increasingly globalised world. It is also about the relationship between humans and the multispecies environment that is holistic rather than compartmentalised, 'and emphasizes the symbiotic character of humans and nature' (Mazzocchi 2006, 464).

Plural, Critical, and Creative Approaches

The case studies in this book, drawn entirely from a European context, cross the boundaries of several ICH domains, including the performing arts. Pauliina Latvala-Harvilahti and Kirsi Laurén's contribution deals with the cultural heritage of mires and asks whether Finnish mire art can potentially change the perception of this cultural heritage. Drawing on two sites in England, Patrick Dillon and Priscilla Trenchard explore contemporary encounters with prehistoric sites through art practice. They examine how artwork is itself an expression of cultural heritage but can also influence the interpretation and management of landscape.

Iain Biggs identifies and explores the implications of creative conversation in relation to two approaches to heritage and change in Wales. Engaging with a complex web of relationships between heritage, inhabitation, mentalities, and ecologies in rural Wales, Biggs argues that creative conversations have important implications for an ecosophical (Guattari 2008) understanding of the relationship between place and heritage. Maarit Grahn examines the impact of nature-based entrepreneurs in Merikarvia, Finland, in maintaining and renewing traditions and values associated with the locality. Discussing intertwined perspectives of commercial, publicly funded, and third-sector agencies, her contribution concentrates on how the values and attitudes businesspeople hold with regard to sustainable development and nature may influence them to become nature-based entrepreneurs. She also examines how features of nature, and local stories associated with these features, are utilised, and what kind of practices entrepreneurs have adopted in order to advance sustainable development.

In approaching these places, and our sense of them, through natureculture and multispecies perspectives, indigenous ways of knowing transcend the confines potentially lurking within the epistemological foundations of these perspectives. Thus, while we may see 'natureculture relation expressed in theories of embodiment that attempt to collapse inside-outside binaries [...] we can see it expressed in examinations of dwelling that are opposed to dichotomizing human-environment relations' (Latimer and Miele 2013, 16). While such dichotomies tend to be associated with 'Western' ways of thinking and knowing that underpin the Capitalocene, technocracy, and colonialism, there are alternative perspectives within those 'Western' ways that have informed creative approaches to engaging with nature. For example, Latimer and Miele (2013, 16) cite Heidegger's (1993) critique of technology, noting that 'humans in modernity increasingly stand in a relation to (not with) nature'. This indicates, quite rightly, that it is not relationality per se that is at stake when we consider new relationships with nature, but both the kind of relationality and its direction. The contributions by Iain Biggs, Mary Modeen, and Ullrich Kockel explicitly follow alternative perspectives that have informed creative approaches, such as Geopoetics, or Creative Ethnology.

Building on research at the interface of human ecology, geography, landscape studies, cultural anthropology, cultural heritage studies, art, archaeology, performance, literature, linguistics, futures research, and philosophy, the contributors to this book address and problematise ways that culture, and cultural heritage, relate to and connect with nature. 'For tribal peoples', Hogan (2014, 19) observes, 'our relationships and kinship with the alive world is simply called tradition'. Examining 'tradition' as a process of intergenerational, interregional, and intercultural transfer of knowledge, skills and values from past to futures, these case studies look at the creative potential of living traditions, living heritage, and Heritage Futures in their socio-ecological contexts, and at the role of ethnologist(s), or other researcher(s), in the processes of tradition and in interdisciplinary research. Concerned with issues such as the spiritual foundations of local community development, or meanings of indigeneity in a post-industrial and post-colonial world, the authors look analytically and hermeneutically at the relationships of human and non-human components of place, and seek out possibilities for sustainable, relational human agency. Their inquiries include: How has the human–nature relationship changed in the current ecological crisis? What alternative kinds of acting in (and with) nature have emerged in the twenty-first century? Are new rules emerging for engaging with natural areas and understanding the communal value of nature? Can art and other media influence the values and attitudes associated with nature? What is the role of heritage practice – including in museums – in the ecological crisis?

Using ethnographic and closely related methods, empirical case studies cover, for example, environmental/political art, film performances, oral narratives, visual images, and media texts, including print, online, and social formats. Located at the interface of ethnology and other arts and humanities fields in an interdisciplinary context, the research underpinning the contributions to this book involves a significant element of practice taking a variety of forms, including audio-visual, fine art, or performance. This diversity of methods and approaches brings with it a variety in the styles of presentation, ranging from more conventional research papers to contemplative essays. In line with the epistemological concerns of the authors, this demonstrates both the potential of individual approaches and their confluence in a shared inquiry.

Sense of Place

Most contributions in this book deal with specific examples in particular places. Grahn examines the phenomenon of nature-based entrepreneurs in Merikarvia, a small rural municipality on the west coast of Finland. Kockel explores museum dynamics in an urban location in Lithuania, while Nic Craith draws her examples from an Irish context. Significantly, not all contributions are land-based. The chapter by Riina Haanpää, Laura Puolamäki, and Eeva Raike focuses on the River Eurajoki in Finland, which runs from Lake Pyhäjärvi to the Bothnian Sea.

Given the significance of place for TEK, this focus on local places is hardly surprising. Kockel (2009, 156) argues that 'all life is taking place in places – where we are. That may sound banal, but should not be forgotten'. The continued presence of a particular group in a specific place legitimates members of a community. A historical narrative lends legitimacy and an air of 'authenticity' to a 'native' culture. Nic Craith (2004) points to the claim to land exemplified in the Welsh language. According to English speakers, people living in Wales are 'Welsh', which is derived from the Anglo-Saxon word *wealh*, meaning 'foreigner'. In the Welsh language, they are the *Cymry*, the 'kinsfolk', who live 'not in the "land of strangers" (Wales), but the "land of fellow countrymen/women" (Cymru)' (Bowie 1993, 186).

Sometimes a focus on the local has generated negative responses. It has led to parochialism, bigotry, and even orchestrated mass murder. And yet, for many people, the physical and cultural connections with place remain significant at many levels, perhaps even more so than events on a global scale. In his poem 'Epic', the Irish poet Patrick Kavanagh contemplates a simple row that took place between two Irish farmers at a time when larger altercations were casting a shadow over Europe on the eve of the Second World War (Nic Craith 2020, xiii). As the Munich Agreement of 1938 ceded the Sudetenland of western Czechoslovakia to Germany, the ghost of Homer came whispering to the poet's ear (Kavanagh 1974, 286):

> He said: I made the Iliad from such
> A local row. Gods make their own importance.

Place is key to understanding who we are and where we come from. It is a key dimension of human being, which is always a 'being there' (Frykman 2003). Our sense of place derives not just from our individual experience but is also framed by our cultural experiences. In *Wisdom Sits in Places*, Keith Basso (1996) explored place, landscape, and language, and what they meant to a particular group of people, the Western Apache in Arizona. In large part, this connection to the land is storied, and therefore made by and through language.

Traditional stories strengthen the connection with the land and include seasonal changes, symbols, and ecological observations. In *Braiding Sweetgrass*, Robin Kimmerer (2020) writes that 'the story of our relationship to the earth is written more truthfully on the land than on the page'. Treating the land as an animate being, she argues that 'the land remembers what we said and what we did'. Kimmerer proposes that we unearth the old stories and generate new stories in order to transform our relationship with nature.

Nic Craith's contribution points to ancient stories in Ireland reflecting a primal understanding of a non-hierarchical relationship between humans and nature. Names of places offer insights into soil properties and animal husbandry. Language gives a sense of rooted-ness. It situates us in our particular environment (Basso 1996). Sometimes such stories can have negative consequences. Latvala-Harvilahti and Laurén's chapter explores how the anomalous nature of mires and peatlands has given rise to mythical and frightening images associated with them. As a result, there has long been a negative perception of pristine mires, and efforts have been made to transform them into something more beneficial to humans.

The meanings of places to people have been and are manifold. People make observations in their living environment and have knowledge related to places. In other words, they have a personal understanding of places, shaped by local, regional, social, cultural, and historical factors. In the chapter by Haanpää, Puolamäki, and Raike, knowledge held by the inhabitants of the River Eurajoki's catchment was primarily of a personal nature, but, of course, included knowledge that had been heard and processed by the community. Above all, it was linked to places that were meaningful to individuals and to the local community, and to which their experiences and stories related. It was local environmental knowledge passed down through generations. The study of the River Eurajoki invites comparison with other river catchment studies that have sought to discern a bioregion by comprehensively locating its natureculture within radical human-ecological time-space (see, e.g., Campbell 2021).

Most of the contributors to this book explore the concept of place. Dillon and Trenchard suggest that landscape is a cultural-ecological artefact which carries within it the imprint of successive generations of human engagement. Modeen points to the growing consensus that 'place' is not a singularly human dimension. Instead, it is co-constituted relationally by a particular environment and human communities. This is achieved in a process both singular and collective, shaped by sensorial embodied experience, and equally by the perceptual interpretation of its character. In each instance the process is framed by language, education, culture, and discourse.

Sited environmental art projects, walking poetry libraries, recovered indigenous narratives, community regeneration projects, and performative 'healings', all point to a growing cognisance of the need to adjust thinking and steward natural balances mindfully. Through a range of 'new nature writing', sited art projects and community actions, specific cases of 'creative knowing' are traced.

Creative Knowing

Several contributions to this volume focus on creatively engaging with nature in order to envisage new nature futures. Nic Craith's contribution explores the pre-Cartesian, pre-Enlightenment perspective on human-nature relations conveyed in the Gaelic languages indigenous to Ireland and Scotland. The rise of rationality and Enlightenment reasoning in Western Europe diminished this thinking and promoted a rational engagement with the Earth that disconnected people and place. Knowing with the senses was regarded as unscientific. Instead, scientists strove towards objectivity. The earth became a resource to be ransacked rather than a partner in a natureculture relationship.

Modeen considers individuals who challenged this way of thinking in a Scottish context, and who regarded the relationship between humans and the natural environment they inhabit as mutually constituent. She points to Nan Shepherd's book *The Living Mountain* (written in 1947 but first published in 1977). Shepherd wrote that one cannot know the rivers till one has seen them at their sources; but this journey to the sources is not to be taken lightly. One walks among elementals, which are not governable. They are awakened also within oneself and are as unpredictable as wind or snow. Shepherd describes the mountain as animate and living. She seeks to know the mountain's essential nature, through first-hand experience and through reflection, appreciative that the 'elementals' in both land and self are ungovernable.

Biggs establishes that creative conversations are the catalyst for a more inclusive and complex sense of the relationships between heritage, inhabitation, mentalities, and ecologies in rural Wales. He suggests that this has important implications for an ecosophical (in Guattari's sense) understanding of the relationship between place and heritage. The first example is the work of Lindsey Colbourne, an artist rooted in Nant Peris in North Wales, whose 'ensemble' practices involve a range of creative approaches oriented by conversation. She has also worked extensively in conflict resolution, public participation, collaboration, and training. A recent interview and the extensive material documented on her website evidence a particular 'conversational aesthetic'. The second approach is that being developed by the Narrating Rural Change network, made up of artists and academics whose point of reference is an arts-sustainability project known as 'Emergence', which has given rise to 'a theory of artfulness as a manner of being that is attuned to a dispersed, emergent version of creativity' (Pigott 2020, 877).

Latvala-Harvilahti and Laurén focus on the meanings that artists attach to mires as natural places, querying how Finnish mire art potentially transforms the cultural heritage associated with mires. As they point out, the phenomenon of environmental art has been ongoing for a long time. Altamira in Spain is famous for its palaeolithic cave art featuring local fauna, estimated to have been sketched around 36,000 years ago. The cave paintings at Lascaux in south-west France, depicting local animals at the time, are estimated to be 20,000 years old.

In the twentieth century, many artists engaged in projects interrogating the more conventional approaches to connecting with nature. An example would be the work of Joseph Beuys, who lived with a coyote in a Manhattan art gallery for three days in 1974. While this may have appeared strange to a conventional audience, it was hardly unusual for Beuys who previously had engaged with a dead hare (Kockel 1995; Walters 2012). van Dooren et al. (2016, 65) suggest that Beuys's legacy has been to nurture new ways of thinking about nature-culture relations among the artistic community: 'Drawing on more than forty years of ecological art that has come since Beuys, contemporary artists are facilitating alternative ways of speaking and thinking about how our own survival is contingent on entanglements within multispecies as assemblages'.

More recently, artist Caroline Wendling piloted the planting of *White Wood*, a living artwork at Hummel Stone in the Bin Forest in Scotland. This involved a group of volunteers planting seven hundred trees, including birch, hazel, rowan, and oak over a weekend. The event in March 2014 was designed to commemorate the centenary of the First World War. The trees, shrubs, and flowers in the wood all have an element of white (the colour of peace) in them. The site itself is:

> a "peace wood"; a monument to the fragility of peace between nations and a memorial both for soldiers from the local area who died on the Western Front and for those who refused to fight and were vilified for their choices.
>
> (Macpherson 2018, 241)

The wood is especially significant as among the trees planted were 49 oak-saplings from the trees planted between 1982 and 1987 for Beuys's 7,000 Oaks project. 'Beuys's environmental art–activist gesture sought to instigate both a global reforestation and a shift in ecological consciousness. Wendling's project knowingly inherits Beuys's desire for regeneration through social sculpture' (Macpherson 2018, 241). While looking back to the calamity of the First World War, Wendling's project was designed with a view towards the future. The survival and 'completion' of the forest is reliant on unidentified volunteers and a society that does not decide to cut down the wood in favour of some factory or building. It is a project that is vulnerable – not unlike the human condition.

Considering more recent artistic engagement with nature, Latvala-Harvilahti and Laurén profile the intensification of the role of artists and art in the current climate crisis with particular reference to the increase in the number of art events and performances taking place on mires (bogs) across Europe in the 2000s. These are ecological actions from artists who want to take a stand for the protection of mires by inspiring people to pay attention to threatened nature. The public are invited to experience nature and art in new ways. In addition, mire-related artists may co-create their art projects with the local residents for whom mires are favourite everyday places.

Dillon and Trenchard explore a similar process with reference to two English sites that are framed as cultural-ecological case studies. Their chapter examines how art can express both contemporary beliefs and multiple meanings associated with the environments of earlier generations. They argue that when framed as cultural ecology, art cannot be understood as separate from the behaviour of the people who produce it, nor from the environmental context in which it is produced, or from the responses of people who engage with it. All are intricately connected. Art is a contribution to the continual re-making of meaning between people and landscape.

Conclusion

Ecological problems are decreasing resilience of our living environments and destroying prerequisites for well-being. Donella Meadows (1999) identified several leverage points for systematic sustainability transformation in society. Of four groups of such leverage points, three focus on political, economic, and institutional spheres of human activities. Arguably, these do not address the fourth group of leverage points – values and worldviews. However, worldviews and values have an impact on all other groups of leverage points, and it has been pointed out that the transformation of values and worldviews is the strongest leverage point for a more sustainable world (Meadows 1999; Abson et al. 2017). This is understandable, because on the one hand, worldviews and values are cultural phenomena present in all human activities, both in private everyday life and in institutional activities, policymaking, business, and administration; on the other hand, it is not possible to direct cultural change by legislation and administration. In the end, culture is unruly (Siivonen 2022).

However, as a part of global sustainability transformation, we first need cultural sustainability transformation. Revised or new understandings of the place of humans in nature, for instance with the help of cultural heritage – especially in the form of Heritage Futures (see, e.g., Kockel 2019) – are key to global sustainability transformation (Siivonen 2022).

Ullrich Kockel's closing contribution in this volume focuses on telling more-than-human stories in and of multispecies urban settings, drawing on the example of a specific initiative by Vilnius City Museum designed to change the hierarchical representation of relations between humans and other species. De-centring the human from the mainstage, a current provisional gallery tells the story of residents of the city from a multispecies perspective. As well as humans, the museum visitor encounters animals, plants, and microbes. Moreover, these are presented in a deep-time perspective and include extinct species. The approach is holistic rather than piecemeal, and the rationale is non-linear. While Western cultures tend to think in linear terms of cause and effect, TEK has often been regarded (e.g., Freeman 1992) as embracing the multi-dimensional nature of reality with its complex web of entanglements. Reflecting on his example, and drawing on the empirical, analytical, and spiritual encounters the other contributors to this book are offering, Kockel ponders ways of cultivating new relationships with nature that overcome the failed binaries by replenishing a deep sense of place as a fundamental component of liveable futures, based on a Humilocene perspective that may point a way forward.

Note

1 Following the style guide of the independent anthropology magazine *Sapiens* (Weeber 2020), we are capitalising 'Indigenous' where, as an adjective, it 'articulates and identifies a group of political and historical communities, compared to the lower case "i", which can refer to anyone.' Where the adjective refers not to such a defined group context, but to aspects of place-based cultural practice, such as 'indigenous language', or where the term is applied more generically, we use the lower case.

Bibliography

Abram, David with Tema Milstein and José Castro-Sotomayor. 2020. "Interbreathing Ecocultural Identities in the Humilocene." In *Routledge Handbook of Ecocultural Identity*, edited by Tema Milstein and José Castro-Sotomayor, 5–25. London: Routledge. https://doi.org/10.4324/9781351068840-1

Abson, David, Joern Fischer, Julia Leventon, Jens Newig, Thomas Schomerus, Ulli Vilsmaier, Henrik von Wehrden, Paivi Abernethy, Christopher Ives, Nicolas Jager and Daniel Lang. 2017. "Leverage Points for Sustainability Transformation." *Ambio* 46: 30–39.

Basso, Keith. 1996. *Wisdom Sits in Places: Notes on Western Apache Landscape.* Albuquerque: University of New Mexico Press.

Berkes, Fikret. 1993. "Traditional Ecological Knowledge in Perspective." In *Traditional Ecological Knowledge: Concept and Cases*, edited by Julian Inglis, 1–9. Ottawa: International Program on Traditional Ecological Knowledge and International Development Research Centre.

Borrows, John and Kent McNeil, ed. 2022. *Voicing Identity: Cultural Appropriation and Indigenous Issues*. Toronto: University of Toronto Press.

Bowie, Fiona. 1993. "Wales from Within: Conflicting Interpretations of Welsh Identity." In *Inside European Identities: Ethnography in Western Europe*, edited by Sharon Macdonald, 167–193. Oxford: Berg.

Campbell, Liam. 2021. *Room for the River: The Foyle River Catchment Landscape: Connecting People, Place and Nature*. Derry: Merdog.

Crutzen, Paul and Eugene Stoermer. 2000. "The Anthropocene." *Global Change Newsletter* 41: 17–18.

Davis, Heather and Zoe Todd. 2017. "On the Importance of a Date, or, Decolonizing the Anthropocene." *ACME: An International Journal for Critical Geographies* 16 (4): 761–780.

van Dooren, Thom, Eben Kirksey and Ursula Münster. 2016. "Multispecies Studies: Cultivating Arts of Attentiveness." *Environmental Humanities* 8 (1): 1–23.

Englert, Sai. 2022. *Settler Colonialism: An Introduction*. London: Pluto.

Freeman, Milton. 1992. "The Nature and Utility of Traditional Ecological Knowledge." *Northern Perspectives* 20 (1): 7–12.

Frykman, Jonas. 2003. "Being there: An Introduction." In *Being There. New Perspectives on Phenomenology and the Analysis of Culture*, edited by Jonas Frykman and Niels Gilje, 7–51. Lund: Nordic Academic Press.

Gadgil, Madhav, Fikret Berkes and Carl Folke. 1993. "Indigenous Knowledge for Biodiversity Conservation." *Ambio* 22 (2/3): 151–156.

Geertz, Clifford. 1957. "Ritual and Social Change: A Javanese Example." *American Anthropologist* 59 (1): 32–54.

Graeber, David and David Wengrow. 2022. *The Dawn of Everything: A New History of Humanity*. London: Penguin.

Guattari, Felix. 2008. *The Three Ecologies*. Translated by Ian Pindar and Paul Sutton. London: Continuum.

Hannerz, Ulf. 1992. *Cultural Complexity. Studies in the Social Organization of Meaning*. New York: Columbia University Press.

Haraway, Donna. 2007. *When Species Meet*. Minneapolis: University of Minnesota Press.

Heidegger, Martin. 1993. "The Question Concerning Technology." In *Martin Heidegger: Basic Writings*, rev. & exp. ed., edited by David Farrell Krell. London: Routledge.

Hogan, Linda. 2014. "We call it tradition." In *The Handbook of Contemporary Animism*, edited by Graham Harvey, 17–26. Abingdon: Routledge.

Iaccarino, Maurizio. 2003. "Science and Culture. Western Science could Learn a Thing or Two from the Way Science is Done in Other Cultures." *EMBO Reports* 4: 220–223.

Ignatiev, Noel. 1995. *How the Irish Became White*. New York: Routledge.

Illich, Ivan. 1971. *Deschooling Society*. New York: Harper & Row.

Illich, Ivan. 1973. *Tools for Conviviality*. New York: Harper & Row.

Jenkins, Willis. 2023. "Ethics after Humanity." *Journal of Religious Ethics* 51 (4): 611–638.

Kavanagh, Patrick. 1974. "Epic." In *The Faber Book of Irish Verse*, edited by John Montague, 286. London: Faber and Faber.

Kimmerer, Robin Wall. 2020. *Braiding Sweetgrass: Indigenous Wisdom, Scientific Knowledge and the Teachings of Plants*. London: Penguin.

Kingsnorth, Paul and Dougald Hine. 2009. *Uncivilisation: The Dark Mountain Manifesto*. Bristol: Dark Mountain.

Kockel, Ullrich. 1995. "The Celtic Quest: Beuys as Hero and Hedge School Master." In *Joseph Beuys: Diverging Critiques*, edited by David D. Thistlewood, 129–147. Liverpool: Liverpool University Press.

Kockel, Ullrich. 2009. "European Ethnology as Engaged Toposophy." *Acta Historica Universitatis Klaipedensis* XIX – *Studia Anthropologica* III: 147–163.

Kockel, Ullrich. 2012. "Being from and Coming To: Outline of an Ethno-Ecological Framework." In *Radical Human Ecology: Intercultural and Indigenous Approaches*, edited by Lewis Williams, Rose Roberts and Alastair McIntosh, 57–71. Aldershot: Ashgate.

Kockel, Ullrich. 2019. "Heritage Futures: Tradition, Gain, Sustainability." Keynote address, conference on *Constructing Social Futures – Sustainability, Responsibility and Power*, University of Turku, Finland, 12–13 June. Accessed 12 September 2024. https://futuresconference2019.wordpress.com/wp-content/uploads/2019/08/ullrich_kockel_heritage-futures-public.pdf

Kockel, Ullrich. 2024. "'Are you (or could you be) Indigenous?' A Perspective from Europe." In *The Routledge International Handbook of Heritage and Politics*, edited by Gönül Bozoğlu, Gary Campbell, Laurajane Smith and Christopher Whitehead, 173–178. London: Routledge.

Kockel, Ullrich and Mairi McFadyen. 2019. "On the Carrying Stream into the European Mountain: Roots and routes of creative (Scottish) ethnology." *ANUAC* 8 (2): 189–211.

Kuokkanen, Rauna. 2000. "Towards an 'Indigenous Paradigm' from a Sami Perspective." *The Canadian Journal of Native Studies* 20 (2): 411–436.

Latimer, Joanna and Mara Miele. 2013. "Naturecultures? Science, Affect and the Non-Human." *Theory, Culture & Society* 30 (7/8): 5–31.

Löfgren, Orvar. 1996. "Linking the Local, the National and the Global: Past and Present Trends in European Ethnology." *Ethnologia Europaea* 26 (2): 157–168.

Macpherson, Alan. 2018. "Art, Trees, and the Enchantment of the Anthropocene: Caroline Wendling's White Wood." *Environmental Humanities* 10 (1): 241–256.

Manemann, Jürgen. 2014. *Kritik des Anthropozäns. Plädoyer für eine neue Humanökologie*. Bielefeld: Transcript.

Mazzocchi, Fulvio. 2006. "Western science and traditional knowledge. Despite their variations, different forms of knowledge can learn from each other." *EMBO Reports* 7: 463–466.

Meadows, Donella. 1999. *Leverage Points: Places to Intervene the System*. Hartland, VT: The Sustainability Institute.

Milstein, Tema and José Castro-Sotomayor. 2020. "Ecocultural Identity: An Introduction." In *Routledge Handbook of Ecocultural Identity*, edited by Tema Milstein and José Castro-Sotomayor, xvii–xxiii. London: Routledge.

Milstein, Tema. 2024. "Re-stor(y)ing the Earth: How language shapes our world." In *Regenerating Our World*. RSA Oceania webinar series, 10 September.

Moore, Jason, ed. 2016. *Anthropocene or Capitalocene? Nature, history, and the crisis of capitalism*. Oakland, CA: PM.

Nic Craith, Mairéad. 2004. "Local Cultures in a Global World." In *Communicating Cultures*, edited by Mairéad Nic Craith and Ullrich Kockel, 279–299. Münster: LIT.

Nic Craith, Mairéad. 2020. *The Vanishing World of the Islandman: Narrative and Nostalgia*. Bern: Springer.

Nic Craith, Mairéad, Ullrich Kockel and Katherine Lloyd. 2019. "The Convention for the Safeguarding of the Intangible Cultural Heritage: Absentees, Objections and Assertions." In *Safeguarding Intangible Heritage: Practices and Politics*, edited by Natsuko Akagawa and Laurajane Smith, 118–132. London: Routledge.

Nic Craith, Mairéad and Philip McDermott. 2025. "Language as Intangible Cultural Heritage: Living Repositories of Knowledge for Tackling Climate Change." *Anthropological Journal of European Cultures* 34 (1): 63–79.

Noyes, Dorothy. 2015. "From Cultural Forms to Policy Objects: Comparison in Scholarship and Policy." *Journal of Folklore Research* 52 (2–3): 299–313.

Pels, Peter. 2022. "'Not an Airy European Fantasy' or How (De)Colonization Embeds the Anthropological Imagination." In *Re-Creating Anthropology: Sociality, Matter, and the Imagination*, edited by David Gellner and Dolores Martinez, 19–33. London: Routledge.

Pigott, Anna. 2020. "Articulating Artfulness: Exploring the Ecological Potential of Creative Conversation." *Transactions of the Institute of British Geographers* 45 (4): 877–890.

Poli, Roberto. 2011. "Ethics and Futures Studies." *International Journal of Management Concepts and Philosophy* 5 (4): 403–410.

Rockström, Johan, Will Steffen, Kevin Noone et al. 2009. "A Safe Operating Space for Humanity." *Nature* 461: 472–475.

Shepherd, Nan. 1977. *The Living Mountain*. Edinburgh: Canongate.

Siivonen, Katriina. 2008. *Saaristoidentiteetit merkkien virtoina. Varsinaissuomalainen arki ja aluekehitystyö globalisaation murroksessa*. Helsinki: Suomen muinaismuistoyhdistys.

Siivonen, Katriina. 2017. "Kulttuurinen kestävyys." In *Kaikki vapaudesta*, edited by Kari Enqvist, Ilari Hetemäki and Teija Tiilikainen, 275–287. Helsinki: Gaudeamus.

Siivonen, Katriina. 2018. "Sustainable Everyday Culture from Glocal Archipelago Culture." In *Cultural Sustainability and the Nature-Culture interface. Livelihoods, Policies, and Methodologies*, edited by Inger Birkeland, Rob Burton, Constanza Parra and Katriina Siivonen, 19–34. London: Routledge.

Siivonen, Katriina. 2022. "Tulevaisuusperintö kulttuurisen kestävyysmurroksen välineenä." In *Tulevaisuudentutkimus tutuksi – perusteita ja menetelmiä*, edited by Hanna-Kaisa Aalto, Katariina Heikkilä, Pasi Keski-Pukkila, Maija Mäki and Markus Pöllänen, 419–433. TVA-julkaisuja 1/2022. Turku: Tulevaisuudentutkimuksen Verkostoakatemia. https://urn.fi/URN:ISBN:978-952-249-563-1

UNESCO. 2003. *Convention for the Safeguarding of the Intangible Cultural Heritage*. https://ich.unesco.org/en/convention

UNESCO (n.d.). *Knowledge and practices concerning nature and the universe*. https://ich.unesco.org/en/knowledge-concerning-nature-00056

Valkeapää, Nils-Aslak. 1983. *Greetings from Lappland: The Sami – Europe's Forgotten People*. London: Zed.

Walters, Victoria. 2012. *Joseph Beuys and the Celtic Wor(l)d: A Language of Healing* Münster: LIT.

Weeber, Christine. 2020. "Why Capitalize 'Indigenous'?" Accessed 7 April 2026. https://www.sapiens.org/language/capitalize-indigenous/

West, Gary. 2012. *Voicing Scotland: Folk, Culture, Nation*. Edinburgh: Luath.

Whyte, Kyle. 2017. "Indigenous Climate Change Studies: Indigenizing Futures, Decolonizing the Anthropocene." *English Language Notes* 55 (1): 153–162.

Wildcat, Daniel. 2009. *Red Alert! Saving the Planet with Indigenous Knowledge.* Wheat Ridge, CO: Fulcrum.

Williams, Lewis, Rose Roberts and Alastair McIntosh, ed. 2012. *Radical Human Ecology: Intercultural and Indigenous Approaches.* Aldershot: Ashgate.

Williams, Lewis, Danielle Alphonse, Ullrich Kockel and Kikila Perrin. 2024. "Transforming Approaches to Reconciliation in Higher Education and Leadership through an Inclusive Practice of Indigenist Social-Ecological Resilience." In *The Bloomsbury Handbook of Ethics of Care in Transformative Leadership in Higher Education*, edited by Mary Drinkwater and Yusef Waghid, 297–315. London: Bloomsbury.

Wulf, Andrea. 2016. *The Invention of Nature: The Adventures of Alexander von Humboldt, the Lost Hero of Science.* London: John Murray.

2 Cultural Sustainability Transformation

The Transformative Role of Heritage Futures

Katriina Siivonen and Ullrich Kockel

Since 1987, when the UN's Brundtland Commission defined sustainable development in terms of 'meeting the needs of the present without compromising the ability of future generations to meet their own needs' (United Nations 1987, 16; 2024), 'sustainability' has acquired various meanings in different contexts, especially in the debate over the myth of sustainable economic growth (Douthwaite 1992). A common core of 'sustainability' can be found in its three principles, or pillars – environmental, social, and economic sustainability (SUSTEMY 2024). In management training programmes, such as RMIT University's online course *Business Futures: Sustainable Business Through Green HR*, these are often expressed as 'The triple bottom line or the three P's [...] People, Planet, and Profit' (RMIT University 2017). Arguably, 'profit' at a steady – or indeed accelerating – rate of growth, regarded as a *sine qua non* from a business perspective, could be considered unsustainable from an Earth Systems perspective. Critics tend to favour 'steady state' (Daly 2014) models of the economy.

During 2005–2014, the UN Decade of Education for Sustainable Development 'aimed at integrating the principles and practices of sustainable development into all aspects of education and learning, to encourage changes in knowledge, values and attitudes with the vision of enabling a more sustainable and just society for all' (UNESCO 2014, 9). One outcome of that Decade has been the establishment of a global network of Regional Centres of Expertise under the auspices of the United Nations University. At the time of writing, some 190 of these RCEs are working worldwide to embed learning for sustainability. Critical of the ambivalences the concept of 'sustainability' entails, steering group members of the RCE *Learning for Sustainability Scotland* have argued that '[i]f sustainability is about imagining and pursuing desired futures (White 2013), our heritage, natural environment and culture will influence the kind of futures we seek and our chosen routes towards these' (White et al. 2023). The different roles of culture and heritage have been discussed in many transdisciplinary contexts, for instance among ca. 100 researchers from 25 European countries and beyond in the project COST Action Investigating Cultural Sustainability in 2011–2015 (Dessein et al. 2015). The present chapter explores new approaches to human-ecology relationships from a systems-based perspective on heritage futures.

DOI: 10.4324/9781003727972-2

Human Impact on Earth Systems

A holistic view on sustainability can be based on an Earth Systems perspective. The Earth Systems, which together provide the foundation for all life, consist of five distinctive systems or spheres – geosphere, atmosphere, hydrosphere, cryosphere, and biosphere – that are dynamically connected to each other, and to human cultural and social systems (see, e.g., National Academies 2022). Human beings are part of Earth Systems both physically, mentally, and culturally. Of crucial importance are the circulation of water, carbon, nitrogen, and phosphorus. These circulations are in a concrete way going through us humans, as through all living beings, when we eat and breathe. We are a part of local and global ecosystems, and our activities have an impact on all the spheres. That impact may be local or regional, as in the case of polluted land areas or water systems, or global, as with the local use of fossil energy that impacts on the global climate. Our shared cultural practices of eating, travelling, dwelling, and acting in general, as well as our worldviews, and the habits through which we make sense of the world, all impact on Earth Systems (Siivonen 2022; Richardson et al. 2023).

The term *Anthropocene* (Crutzen and Stoermer 2000) seeks to capture the alteration of Earth Systems through human agency in terms of a new geological epoch. To date, there is no officially accepted way of measuring human impact on biophysical processes, and the debate over the concept remains ongoing, both in geology and beyond. However, human impact on Earth in our era is clear (Lewis and Maslin 2015; Johnson et al. 2022).

Different interconnected systems of human impact in relation to planetary boundaries have been analysed in meta-analyses in natural sciences (see Figure 2.1). The aim of these analyses is to gather scientific evidence of the level of human impact on Earth Systems, and to identify critical processes for maintaining the stability and resilience of these systems as a whole. In these cumulative analyses, six elements have at the moment appeared to be at a high risk level. They are: two elements belonging to the loss of biodiversity (genetic and functional biosphere integrity); two elements of biochemical flows, connected to, for example, farming including fertilisation (the circulation of nitrogen and phosphorus); climate change; and, novel factors, which includes synthetic chemicals, microplastics, nuclear waste, and human interventions in evolutionary processes. Land-system and freshwater change is identified as posing an increasing risk (Rockström et al. 2009; Steffen et al. 2015; Richardson et al. 2023).

In the current era, there is a need to combine natural scientific knowledge on nature and humanistic knowledge of humans and nature. There has been growing recognition of the economy as culturally contingent; and so is the discipline purporting to explain it, economics (Gudeman 2001), with its insistence on the profit motive as a pillar of sustainability. This has highlighted the need for complementary perspectives to integrate the emerging scientific view of humans and culture as a part of nature. Critical analysis of cultural and environmental policies should rely, in the first instance, on the

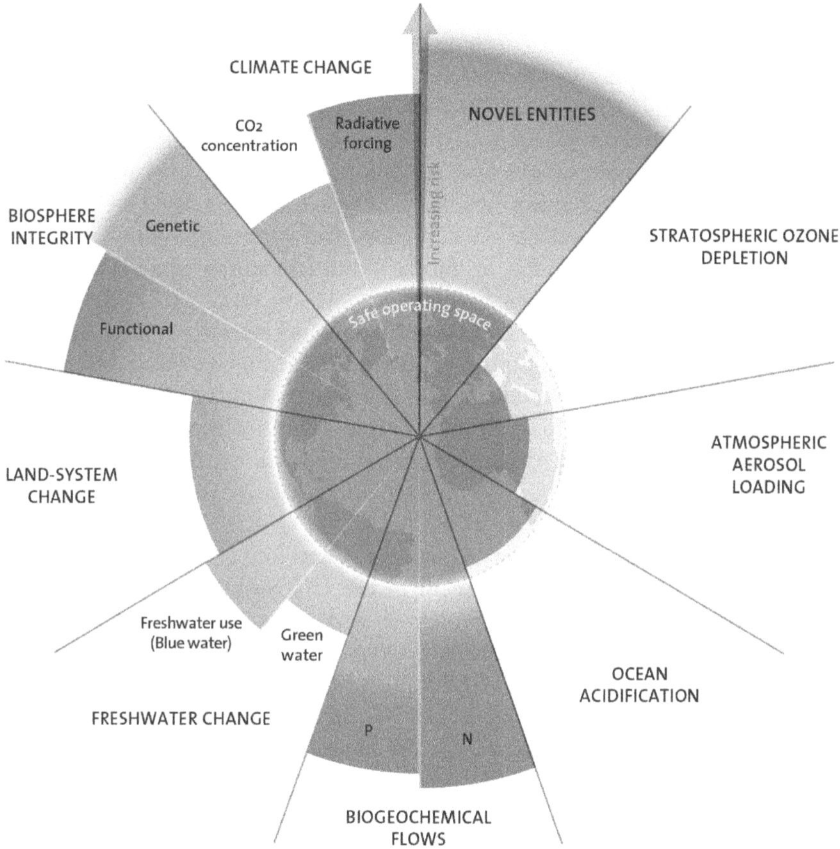

Figure 2.1 Azote for Stockholm Resilience Centre, based on analysis in Richardson et al. (2023). Attribution: CC BY-NC-ND 3.0. The printed book uses a black and white version of the original colour figure, used by permission.

systematic scientific analysis of interconnections between the human species and its communities and culture in their respective glocal bioregional setting. Combining a Cartesian approach with other ways of knowing may lead us to new ways of engaging in natureculture (Haraway 2007) relationships. Drawing, for instance, on the recovery of indigenous wisdom that is contained within traditional ecological knowledge (Nic Craith, in this volume), a vernacular understanding of the place of humans in nature might be developed that will hone our abilities to transform our culture according to its limits within planetary boundaries.

The Need for Cultural Transformation

Among others, Siivonen et al. (2024) point out that, considering the alarming human impacts on nature, it is obvious that human behaviour must change if the resilience of Earth Systems is to be maintained and a more sustainable world reached. For some decades, leverage points for systemic sustainability transformation (see, e.g., Meadows 1999) have been understood. Leverage points target the most effective spots in a society in order to enhance sustainability transformation. They are categorised on four levels (Abson et al. 2017): political, administrative, and economic actions; practices of organisations and institutions; interaction among organisations and institutions; and worldviews and values.

For successful sustainability transformation, there is a need to reach the deepest level of leverage points – worldviews and values, including an understanding of the role of humans in nature – and to identify goals and actions arising from these. It has been pointed out (Abson et al. 2017) that policy interventions typically do not address the deep level leverage points, and thus fail in their target of sustainability transformation, but worldviews and values have an impact on all other levels of leverage points (Meadows 1999). Multiple worldviews, values, human-nature relationships, and goals and actions arising from them, are basic cultural expressions. As cultural expressions, they are defined by human private and professional life, and they in turn define human life (Siivonen et al. 2024).

It is worth noting that deep level leverage points and cultural sustainability transformation are not targeted at all in the UN Sustainable Development Goals (United Nations 2023), although they have an impact on each one of them. As part of sustainability transformation at a local as well as a global level, there is a need for goal-oriented cultural transformation (Siivonen 2022; see also Huttunen et al. 2021). Worldviews, values, human–nature relationships and related knowledge, skills, actions, habits, and practices belong to intangible culture. In instrumental use in cultural, social, economic, and political contexts – for example, in UNESCO conventions – these are defined as Intangible Cultural Heritage (ICH). Culture and cultural heritage have an impact everywhere in society.

Conceptual Torches

Culture is a central driving force of futures and potentially also towards a more sustainable world (Siivonen 2017). It is therefore necessary to define it in the context of cultural sustainability transformation. Definitions of culture abound in cultural research and in cultural policy (Wiktor-Mach 2018). Concepts are like torches: In the beam of light of one definition of a concept, we can see certain issues, while others are left in the shadow. This directs our understanding of cultural phenomena, including related actions. We need to be aware of different definitions of operational terms and their capabilities to shed light on the world. It is important to assess which definitions are relevant in different contexts in order to enable both practical societal actions relating to, and scientific discussion about them. Clear and unambiguous definitions are needed, especially in contexts where there is a rich variation of meanings and a term is in vernacular use, as is the case with the concept of 'culture' (Siivonen 2022).

According to an ethnological definition, culture is at least something tangible and intangible that humans share with one another (Gerholm 1994), covering the way human beings are in the world together with other human beings. It consists of worldviews and values with their impact on actions, practices, habits, and the material world, which we all create, co-create, receive, learn, teach, reject, and change in interaction with others. Culture covers all spheres of both private everyday actions and the professional life of humans: economy, technology, policy making, administration, education, built environments, art, ecological relationships, and so on (see, e.g., Siivonen 2008a, 2008b). In cultural policy and related research, the main focus of the concept of culture may be on art, cultural heritage, and cultural production (see, e.g., Wiktor-Mach 2018). Depending on the context, both ethnological and cultural policy definitions of culture can be relevant in relation to futures and sustainability, helping us understand how cultural products and cultural production, for instance cultural heritage, are used as tools to control change and transformation in human life (Siivonen et al. 2024).

According to Goody (1994), definitions of culture used in anthropology or ethnology can be categorised into two basically different types, which can be characterised as a framework and a river. As a framework, culture is seen as a relatively stable structure defining social activities (Siivonen 2008a, 2008b). This is following the definition of culture by Geertz (1957, 33), who suggested that culture is 'the framework of beliefs, expressive symbols, and values in terms of which individuals define their world, express their feelings, and make their judgements'. The focus is on the character of culture and associated worldviews as markers of human beings, and of their relationships with other human beings and their surroundings. 'Culture as a river' indicates a social process in which culture is constantly renewed and recreated in interaction between individual human beings, and between humans and their surroundings. This global process, and its temporal determination of perceived structures, has been highlighted by Ulf Hannerz (1992, 4):

> When you see a river from afar, it may look like a blue (or green, or brown) line across a landscape; something of awesome permanence. But at the same time, 'you cannot step into same river twice,' for it is always moving, and only in this way does it achieve its durability. The same way with culture – even as you perceive structure, it is entirely dependent on ongoing process.

The focus here is on actions, skills, practices, and habits arising constantly from value-based worldviews, and changing in the stream of the cultural river (Siivonen 2008a, 2008b). As it is not possible to stop it from changing thus, culture has an elementary transformative power, but without any common goal-oriented direction (Siivonen 2017). From the perspective of cultural sustainability transformation, it should be noted that every human is always, consciously or otherwise, making ethical, value-based choices directed towards all forms of life, now and in the future, and this adds individual goal-orientations to cultural change (Siivonen 2022; Poli 2011).

The different definitions of culture may shed light on issues relevant in the context of futures and cultural sustainability transformation. However, with the definition of culture as river it is possible to focus on the role and agency of individual human beings in the 'carrying stream' (Hamish Henderson; see Kockel and McFadyen 2019) of cultural tradition. Conscious transformation in values and worldviews is possible only through human agency, even though, due to various, partly contradictory agencies, it is not possible to direct the transformation as a whole (Siivonen 2017).

Anthroposemiosis

We can approach the concept of 'culture as river' more closely by using semiotics and the concept of semiosis. Every human observes her or his individual environment, interprets it, and acts in it, in a personal manner with the help of signs, in continuous interaction with other people and their shared environment (Siivonen 2008a, 2008b). According to the semiotics of Charles S. Peirce (see Figure 2.2), every sign is a triadic relation of, firstly, the *interpretant*, that is, the sign itself, as perceived in the human mind; secondly, the *representamen*, the material manifestation of the sign; and thirdly, the *object* of the sign (CP: 2.228, 274–302;[1] Deely 1994; Siivonen 2008a, 2008b). For example, the exceptional floods in Greece in the summer of 2023 are a material manifestation (*representamen*), and knowledge about the increasing risk of floods due to climate change is an *object* of the sign, which combined allow these floods to be seen as a sign (*interpretant*) of climate change.

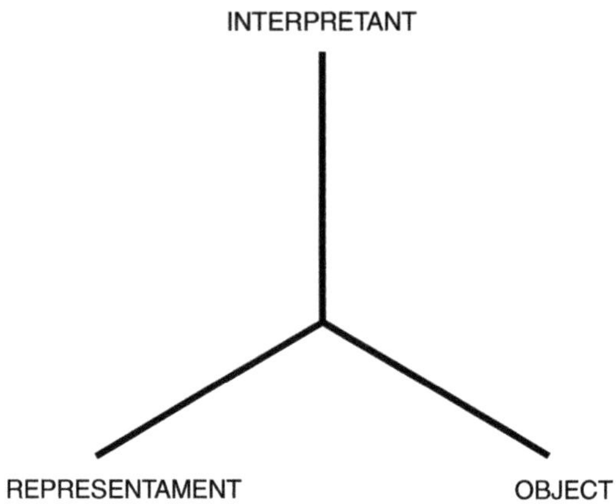

Figure 2.2 Triadic sign of Charles S. Peirce (CP:2.228, 274-302), a relation between representamen, object, and interpretant.

Source: Graphic after Merrell (1995, 41–43); Siivonen (2008a, 50).

In daily life, in semiosis, we use many different signs (Deely 1994; Siivonen 2008a, 2008b). The example of floods in Greece illustrates vague, abductive knowledge: it might be so, even probable, but in our everyday life we might not have exact scientific evidence of it. However, this kind of abductive, cultural knowledge we use constantly in our daily life (Siivonen 1994). Abductive knowledge can also be used in creative processes and, for instance, in the imagining of alternative futures both in private life and in professional contexts (Patokorpi and Ahvenainen 2009; Siivonen 2022). Signs can be conscious symbols, as in the example of the floods in Greece. They can also be unconscious habitual signs or sense-based impressions, and act as tacit knowledge (Merrell 1995) in interaction processes. As an example, without conscious reasoning, we interpret many matters about other people on the basis of their outfit (Siivonen 1994, 2008a, 2008b).

In interaction, signs are connected to other signs, forming chains that intertwine with each other (see Figure 2.3). This is the process of semiosis

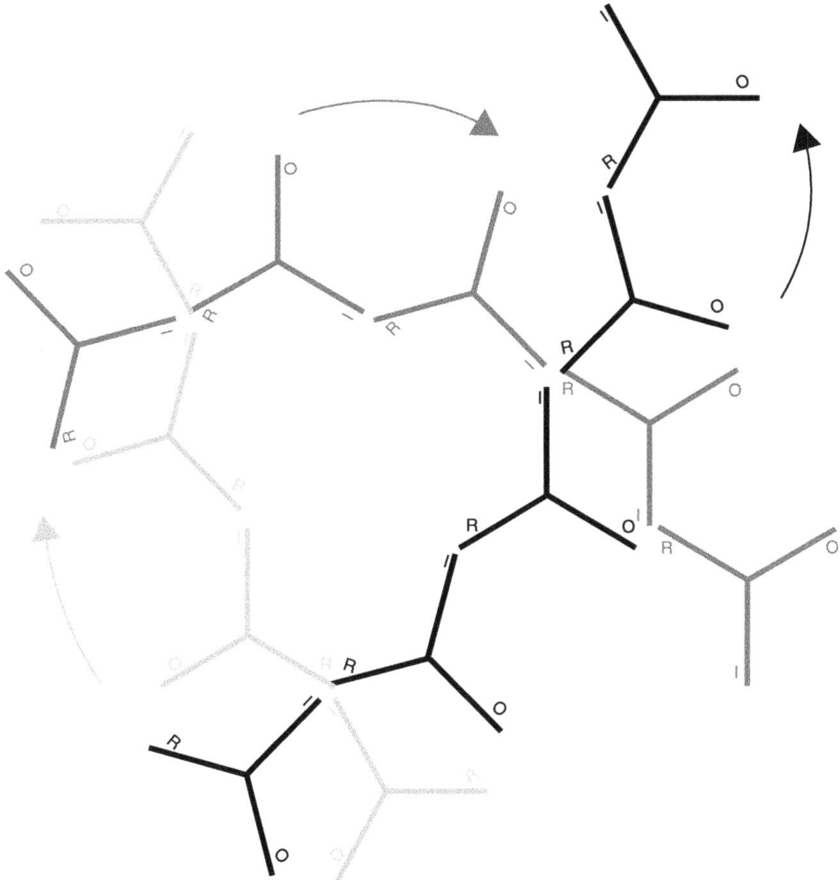

Figure 2.3 Intertwined chains of signs in semiosis. *R* = representamen, *O* = object of the sign, and *I* = interpretant (Siivonen 2008a, 53).

that covers the whole universe (CP: 2.228; CP 7.536, 587; Deely 1990, 1994). When human beings are involved, this becomes anthroposemiosis (Deely 1990, 1994). Following Siivonen (2008a, 2008b), the river of culture can be defined as anthroposemiosis. In this process of anthroposemiosis, the primary form of existence of culture is as a global network or process of signs. While it is impossible to find any part of it that would be perceived by different people in exactly the same way, nobody can be outwith the structure or process. In anthroposemiosis, signs are seen in a state of inevitable change, with no sign following another ever being identical to the previous one (Siivonen 2008a, 2008b; Bergman 2004, 401–402).

Anthroposemiosis is the primary power in culture, a driving force of futures and possible driving force towards a more sustainable world in the global river of culture. In the globalised world, global and local elements are constantly intertwined in glocal, heterogeneous, and dynamic entities. The basic quality of culture is, on the one hand, the individual dynamic and creativity, and, on the other hand, the individual capacity to build on the foundation of different prior experiences, actions, habits, and symbols that are shared with other people in the glocal process of anthroposemiosis. In the current course of globalisation, organisational power is no longer able to form and uphold extensive and homogeneous organisational and cultural wholes. Different competing symbolic worlds will become more common, increasing in importance in global, local, and glocal processes (Siivonen 2008b).

Intertwined Matter and Meanings

To understand how matter and meaning intertwine with each other in time, Siivonen (2021) compiles concepts of memories and images of futures, as well as semiosis and anticipation (Poli 2017). Memories (Fingerroos and Haanpää 2006) and images of futures (Rubin 2000) are inevitably current cultural phenomena, even though they refer beyond the present moment to past times and futures. The current situation of individuals and communities has an impact on what we now think about past times, what we remember, and how we understand futures. The worldviews and values individuals in the present hold and share with others, as well as practical, contextual experiences, wishes, hopes, fears and so on, give different meanings to both memories and images of futures. They change over time and vary according to different contexts and situations (see, e.g., Rubin 2000; Fingerroos and Haanpää 2006). Individual and collective memories and images of futures have an impact on those futures (Rubin 2000; see also, e.g., de Jouvenel 1967). That impact is not direct; it happens in the structural framework or the carrying stream of cultural tradition. In our current world with complex and global interdependencies, different cultural frameworks meet tensions, and cultural rivers are streaming strongly (Siivonen 2021, 2022).

In anthroposemiosis, individual and collective memories and images of futures can be defined as dynamic signs. As such, they are in a state of inevitable change (Bergman 2004, 401–402). The change, as well as resistance to it, is produced by human beings. Memories and images of futures can be seen as including conscious or unconscious anticipatory perceptions and interpretations connected with human actions and practices, including at least implicit intentionality towards futures. Thus, culture and, as a part of it, memories and images of futures, always have, at least implicitly, transformative power (Siivonen 2017, 2021; Poli 2017, 2–6).

Anthroposemiosis can be related to human anticipatory communication and actions, which are part of a broader sphere, and tightly intertwined with the anticipatory communication of other species, with inorganic nature and with other non-living beings in common semiosis (Siivonen 2021; Deely 1994, passim; Poli 2017, 2–3, 5).

Through the relations between tangible and intangible parts of the triadic sign, semiosis – including anthroposemiosis – is a continuous process of interconnected flows of matter and meaning. Material flows consist of changing matter and energy, including embodied human beings in their everyday interactions. Flows of meaning in semiosis consist of changing conscious and unconscious perceptions, worldviews, values, memories, images of futures, actions, habits, practices, skills, and stories that people, and partly also other living beings, share with each other. Culture is constantly recreated in interaction between matter and meaning. These two flows intertwine with each other every moment in individual human actions in the river of culture, or in global anthroposemiosis. Each of us is involved in the carrying stream of culture in our own private and professional life (Siivonen 2021, 2022; Barad 2003; see also Siivonen 2008a).

In anthroposemiosis, the material world becomes part of human understanding, and human beings have an impact on their material surroundings (Deely 1994; Siivonen 2008a, 2008b). If we see embodied human beings as a part of the planetary material flows, anthroposemiosis could be seen, following Karen Barad (2003), as part of the universe of agency continuously (re)articulating semantic material-discursive practices (Siivonen 2021).

The Human Part of Semiosis and Anticipation

In the era described as the *Anthropocene*, with clear human impact on the stability and resilience of Earth systems, and when defining culture as an interactive part of nature (e.g., Siivonen 2018; Willamo et al. 2017), it is possible to notice that anthroposemiosis seems to be a dominant process in semiosis. The relationship between humans and other parts of nature can be analysed with the help of concepts of cultivation and culturalisation. In cultivation, yield is a product of human activities, but it is in part shaped by nature, and thus outside the direct control of culture. In culturalised products, such as meals, furniture, mobile phones, produced energy, and so on, people use natural resources, but the form of the product is shaped according to cultural traditions (Simmel 1988; Siivonen 2018).

In anthroposemiosis, when humans are involved in semiosis, nature in its cultivated and culturalised form is a part of signs in the human mind (Deely 1994, 22–31; Siivonen 2018). Within the South–West Finnish Archipelago identities and culture, an undefined part of the sea is considered unknowable, and thus impossible to control with cultivation and culturalisation, but its presence as such is still known. In this example, the sea is explicitly a part of a larger semiosis that covers the whole physical and symbolic world, and includes anthroposemiosis. This is the case in regions where human actions are tightly interconnected in nature (Siivonen 2018; Keitumetse 2011).

The Right to Enhance Transformation

In order to reach ecological sustainability, an anticipatory understanding of semiosis inside and outside of anthroposemiosis is needed. Firstly, the construction of an ecologically sustainable world requires changes in cultural expressions (Siivonen 2022). Secondly (Siivonen 2022), we must respect individual worldviews according to, for instance, the UN Human Rights Declaration (United Nations 1948): 'Everyone is entitled to all the rights and freedoms set forth in this Declaration, without distinction of any kind, such as [...] political or other opinion'.

It is important to reflect on who has ownership of (cultural) sustainability transformation (Siivonen 2022). The ethical responsibility in scientific work on futures, according to professional ethics in futures research, demands assistance for people to understand how different values impact on different futures (Poli 2011; Siivonen 2022). The ethical responsibility in scientific work on cultures in the professional ethics of anthropology and ethnology demands evaluation of the impact of the research processes on individuals, communities, and environments as well as on the forms of cultural expressions and cultural rights in relation to them (American Anthropological Association 2023; Siivonen 2022).

These expressions are, for instance, worldviews and political or other opinions as well as different practices and habits based on them. When combining cultural heritage research, futures research, and sustainability science for the co-creation of cultural sustainability transformation, there is a need to find a balance between different ethical standpoints on these (Siivonen 2022). Such ethical perspectives suggest, firstly, a broad inclusion and participation of people in transformation processes. From this perspective, the core of cultural sustainability is the right of people to take part in, and have an impact on, their own culture and any intentional changes to it achieved together with other people. However, this kind of process does not necessarily lead to a more sustainable world. As mentioned, they can include contradictory economic, social, ecological, and cultural value-based changes, and cultural transformation may have various directions (Siivonen 2017, 2022).

Paradoxically, voluntary participation and individual, intentional changes are nevertheless necessary in cultural sustainability transformation, which is not possible without them. Hence, we need to define methods for the voluntary, conscious cultural transformation that is regarded as necessary to tackle the ecological crisis (Siivonen 2022, 2017). This kind of dynamic, voluntary grassroots perspective is hard to combine with the implementation of the UN Sustainable Development Goals as global, regulated actions.

Sustainability and Heritage

When we define cultural sustainability within the framework of culture, it is usually with a view to the preservation and safeguarding of cultures, and cultural elements belonging to these, where culture is understood in structural terms. Cultural sustainability in the carrying stream of culture is a comprehensive, anticipatory, social process of tradition (Kockel 2019), in which culture is constantly renewed and co-created in interaction between human beings, and between human beings and their tangible and intangible, cultivated and non-cultivated as well as culturalised and non-culturalised surroundings. Cultural sustainability, then, is the safeguarding of the living process and the transformative power of culture (Siivonen 2022, 2017, 2008a).

Cultural heritage comprises elements of intangible and/or tangible cultural phenomena selected from the river of culture to be used as an instrument to impact on futures (Kockel 2019; Siivonen 2017; Siivonen et al. 2024). Both in everyday life and in cultural policy, people select elements important for themselves as cultural heritage, either to safeguard or preserve them – or to remember them as something to avoid in the future, which is then defined as dark heritage (Siivonen et al. 2024; Thomas et al. 2019). Material items, buildings, and landscapes have long been designated for preservation as tangible heritage or antiquities. There are established institutions and legal instruments for the preservation of heritage, both nationally and internationally. Preserving and safeguarding of cultural elements have directed the focus of heritagisation processes to the past. However, cultural heritage has always targeted futures, and its impact permeates society. The focus has currently turned to communities and their ways of maintaining and developing their intangible cultural heritages, sometimes in the form of institutionalised ICH, which includes knowledge, skills, habits, and everyday practices (Smith 2006; Siivonen 2017, 2019; Siivonen et al. 2024; see also, e.g., UNESCO 1972, 2003).

As outlined by Siivonen (2019), legally defined ICH belongs to the current era of global transformation. We are no longer just acting culturally – we are aware of our actions as cultural phenomena. The global mobility of contemporary human beings strengthens this awareness. In heritagisation processes, the intention is to decrease the power of experts and give space for individuals and communities to define their own heritage (Ahmas 2014). Different communities are increasingly seeking to impact their environment and future by drawing on culture and heritage as instruments. Thus, heritagisation and intentional use of culture expand into societal contexts where they have not been used earlier, which changes how we consider cultural phenomena (Siivonen 2019).

With the help of cultural policy, and actions in the society deriving from this, it is possible to enhance the well-being of humans and nature, and intergenerational and interspecies justice. People can empower each other to create novel intangible cultural heritage, also as official ICH, aiming to decrease the exploitation of natural resources by curbing overconsumption. The use of novel heritage can, if we want, lead to cultural sustainability transformation. There are different forms of transformative heritage processes. Some of them can support resilience and cohesion of individuals and communities or raise awareness of social or ecological problems. With other forms, it is possible to target intentional change in human behaviour (Siivonen 2022).

When evaluating different kinds of impacts of artistic and heritage actions, it is probable that environmental art can raise awareness of the need for sustainability transformation but does not directly produce cultural transformation. Furthermore, safeguarding of cultural heritage can support resilience of communities and individuals in relation to social challenges and environmental crisis, which can support cultural sustainability transformation, but does not directly produce it either (Siivonen 2022; see Newell et al. 2017). Co-creation of Heritage Futures (Kockel 2002; Siivonen 2022), based on new forms of intangible cultural heritage and official ICH, can be one way to produce novel heritage and accelerate cultural sustainability transformation. Utilising Heritage Futures requires support for understanding cultural change in the past, present, and future, and intention to recognise and find solutions to problematic cultural behaviour.

Heritage Futures

Working with an international research consortium, Rodney Harrison developed a concept of Heritage Futures that relies on the idea of managing the preservation of tangible cultural heritage and its consequences for human behaviour and well-being, as well as for intangible heritage (Harrison et al. 2020). We suggest a different concept of Heritage Futures that may be a tool to co-create alternative, sustainable futures, and accordingly actions in culture and society. Siivonen's definition of the concept is based on intangible cultural heritage, in other words, worldviews, values, skills, habits, and practices, and culture seen as an anticipatory process of anthroposemiosis. With this, it is possible to focus on human beings as proactive actors in culture, society, and nature and, if they so wish, voluntary promoters of cultural sustainability transformation (Siivonen 2019, 2022). Kockel initially defined Heritage Futures as expectations heritage processes create in form of human activities affecting the future of human beings, communities, culture and nature (Kockel 2002), and critically compared the creation of novel heritage to economic futures markets, raising the issue of sustainability of such cultural innovation, and proposing 'gain' as a more appropriate concept than 'profit' for measuring the success of Heritage Futures (Kockel 2019). Together, both viewpoints form an important element of the concept of Heritage Futures.

In order to help us better to understand the human–nature interface, and to produce practical sustainable solutions to the current crisis, Heritage Futures needs to include both anticipatory understanding of futures, and an understanding of semiosis beyond mere anthroposemiosis (Kockel 2019; Siivonen 2022). Comprehension of nature and cosmos beyond human perception can support respect for nature as living environment for humans and other living beings, not just as resource and ecosystem services to be exploited by humans. Experience of both culturalised, cultivated, and especially un-cultivated nature is an elementary part of this (Siivonen 2018; Boudes 2011; Simmel 1988). The basic idea of Heritage Futures is as an inspiring, affectual, intentional, anticipatory, cultural tool to co-create alternative images for sustainable futures, and accordingly skills and actions in culture and society (Siivonen 2022).

To be culturally created and strengthened, with respect for ethical standpoints, voluntary participation is a prerequisite for the implementation of cultural sustainability transformation. Heritage Futures is developed as an instrument for cultural sustainability transformation that combines participatory ethnographic and futures studies methodology offering possibilities to find solutions and hope together. In Heritage Futures Workshops, people can, in a voluntary process, co-create with each other ways to enhance cultural sustainability transformation. In other words, by respecting planetary boundaries, they define new or renewed worldviews, values, and skills that can together be recognised and acknowledged as Heritage Futures. As a heritage process, it might take time for Heritage Futures to be rooted in the local and glocal, place-based streams of tradition (Siivonen 2022; Paaskoski et al. 2025).

Heritage Futures Workshops belong to participatory futures research methods. They are based on the methodology of Futures Workshops, with an addition of conceptual understanding of intangible cultural heritage, official ICH and Heritage Futures as well as ethnographic and other qualitative methodology (Paaskoski et al. 2025). Futures workshops are organised as empowering processes in every detail to support co-creation and the possibility of participants to co-imagine futures, have an impact on the change or transformation of their own culture, and gain ownership of their own future (Jungk and Müllert 1987; Armanto et al. 2022). Heritage Futures Workshops are developed with museums wishing to involve users of their collections and displays in the co-creation of sustainable futures, but can also be used in other contexts (Paaskoski 2020; Paaskoski et al. 2025).

Heritage Futures require disruptive co-creation of selected cultural elements and care of their sustainable impacts on nature and human beings, alongside the safeguarding of such selected cultural elements, with respect for planetary boundaries. Care of the transformative power of culture is a part of Heritage Futures. Heritage Futures implies intentionally co-created understandings of the place of humans in nature, including new types of meanings and actions with which humans should be able to produce cultural sustainability transformation (Siivonen 2022; Siivonen et al. 2022).

Note

1 Following the accepted practice, the Collected Papers of Charles S. Peirce are referenced as CP followed by volume and paragraph numbers, separated by a full-stop.

Bibliography

Abson, David J., Joern Fischer, Julia Leventon, Jens Newig, Thomas Schomerus, Ulli Vilsmaier, Henrik von Wehrden, Paivi Abernethy, Christopher D. Ives, Nicolas Jager and Daniel J. Lang. 2017. "Leverage Points for Sustainability Transformation." *Ambio* 46: 30–39. https://doi.org/10.1007/s13280-016-0800-y

Ahmas, Kristiina. 2014. *Norsunluutornin purkajat. Jaettu johtajuus ja kollektiivinen asiantuntijuus museossa.* Vaasa: University of Vaasa.

American Anthropological Association. 2023. *AAA Statement on Ethics, Principles of Professional Responsibility.* https://americananthro.org/about/anthropological-ethics/

Armanto, Riikka, Ville Lauttamäki and Katriina Siivonen. 2022. "Monimuotoinen tulevaisuusverstas." In *Tulevaisuudentutkimus tutuksi – perusteita ja menetelmiä*, edited by Hanna-Kaisa Aalto, Katariina Heikkilä, Pasi Keski-Pukkila, Maija Mäki and Markus Pöllänen. TVA-julkaisuja 1/2022, 222–236. Turku: Tulevaisuudentutkimuksen Verkostoakatemia. https://urn.fi/URN:ISBN:978-952-249-563-1

Barad, Karen. 2003. "Posthumanist Performativity: Toward an Understanding of How Matter Comes to Matter." *Sign: Journal of Women in Culture and Society* 28 (3): 801–831.

Bergman, Mats. 2004. *Fields of Signification. Explorations in Charles S. Peirce's Theory of Signs.* Philosophical Studies from the University of Helsinki 6. Helsinki: University of Helsinki.

Boudes, Philippe. 2011. "La nature au secours de la culture? Lecture simmelienne de la végétalisation des villes et de leurs fonctions." *Revue d'Allemagne et des Payes de Langue Allemande* 43 (1): 131–146.

CP: Peirce, *Charles Sanders. 1931-58. The collected papers of Charles Sanders Peirce.* edited by Charles Hartshorne and Paul Weiss (volumes 1–6, published 1931-35), and Arthur W. Burks (volumes 7-8, published 1958). Cambridge/MA: Harvard University Press.

Crutzen, Paul and Eugene F. Stoermer. 2000. "The 'Anthropocene.'" *IGBP Newsletter* 41.

Daly, Herman. 2014. *From Uneconomic Growth to a Steady-State Economy.* Cheltenham: Elgar.

Deely, John. 1990. *Basics of Semiotics. Advances in Semiotics.* Bloomington and Indianapolis: Indiana University Press.

Deely, John. 1994. *The Human Use of Signs or Elements of Anthroposemiosis.* London: Rowman & Littlefield.

Dessein, Joost, Katriina Soini, Graham Fairclough and Lummina Horlings, ed. 2015. *Culture in, for and as Sustainable Development. Conclusions from the COST Action IS1007 Investigating Cultural Sustainability*. Jyväskylä: University of Jyväskylä. https://jyx.jyu.fi/handle/123456789/50452

Douthwaite, Richard. 1992. *The Growth Illusion. How Economic Growth Has Enriched the Few, Impoverished the Many and Endangered the Planet*. Dartington: Green Books.

Fingerroos, Outi and Riina Haanpää. 2006. "Muistitietotutkimuksen ydinkysymyksiä." In *Muistitietotutkimus. Metodologisia kysymyksiä*, edited by Outi Fingerroos, Riina Haanpää, Anne Heimo and Ulla-Maija Peltonen, 25–48. Helsinki: Finnish Literature Society.

Geertz, Clifford. 1957. "Ritual and Social Change: A Javanese Example." *American Anthropologist* 59 (1): 32–54.

Gerholm, Lena. 1994. "The Concept of Culture in Ethnology." In *Culture and Management in a Changing Europe*, edited by Annick Sjögren and Lena Janson, 15–28. Botkyrka: Multicultural Centre and Stockholm: Institute of International Business.

Goody, Jack. 1994. "Culture and Its Boundaries: A European View." In *Assessing Cultural Anthropology*, edited by Robert Borofsky, 250–261. New York: McGraw-Hill.

Gudeman, Stephen. 2001. *The Anthropology of Economy. Community, Market, and Culture*. Oxford: Blackwell.

Hannerz, Ulf. 1992. *Cultural Complexity. Studies in the Social Organization of Meaning*. New York: Columbia University Press.

Haraway, Donna. 2007. *When Species Meet*. Minneapolis: University of Minnesota Press.

Harrison, Rodney, Caitlin DeSilvey, Cornelius Holtorf et al. 2020. *Heritage Futures, Comparative Approaches and Cultural Heritage Practices*. London: UCL Press.

Huttunen, Suvi, Minna Kaljonen, Annika Lonkila et al. 2021. "Pluralising Agency to Understand lBehaviour Change in Sustainable Transitions." *Energy Research & Social Science* 76. https://doi.org/10.1016/j.erss.2021.102067

Johnson, Amy, Chris Hebdon, Paul Burow, Deepti Chatti and Michael Dove. 2022. "Anthropocene." In *Oxford Research Encyclopedia of Anthropology*, edited by Mark Aldenderfer. Oxford: Oxford University Press. https://doi.org/10.1093/acrefore/9780190854584.013.295

Jouvenel, Bertrand de. 1967. *The Art of Conjecture*. New York: Basic Books.

Jungk, Robert and Norbert Müllert. 1987. *Future Workshops, How to Create Desirable Futures*. London: Institute for Social Inventions.

Keitumetse, Susan O. 2011. "Sustainable Development and Cultural Heritage Management in Botswana: Towards Sustainable Communities." *Sustainable Development* 19 (1): 49–59. https://doi.org/10.1002/sd.419

Kockel, Ullrich. 2002. "Culture and Economy: A Brief Introduction." In *Culture and Economy. Contemporary Perspectives*, edited by Ullrich Kockel, 1–10. Aldershot: Ashgate.

Kockel, Ullrich. 2019. "Heritage Futures: Tradition, Gain, Sustainability." Keynote address, conference on *Constructing Social Futures – Sustainability, Responsibility and Power*. University of Turku, Finland, 12–13 June. Accessed 12 September 2024. https://futuresconference2019.wordpress.com/wp-content/uploads/2019/08/ullrich_kockel_heritage-futures-public.pdf

Kockel, Ullrich and Mairi McFadyen. 2019. "On the Carrying Stream into the European Mountain: Roots and routes of creative (Scottish) ethnology." *ANUAC* 8 (2): 189–211.

Lewis, Simon L. and Mark A. Maslin. 2015. "Defining the Anthropocene." *Nature* 519: 171–180. https://doi.org/10.1038/nature14258

Meadows, Donella. 1999. *Leverage Points: Places to Intervene in a System*. Hartland: The Sustainability Institute.

Merrel, Floud. 1995. *Peirce's Semiotic Now. A Primer*. Toronto: Canadian Scholars' Press Inc.

National Academies of Science, Engineering, and Medicine. 2022. *Next Generation Earth Systems Science at the National Science Foundation*. Washington, DC: The National Academies Press. https://doi.org/10.17226/26042

Newell, Jennifer, Libby Robin and Kirsten Wehner. 2017. "Introduction: Curating Connections in a Climate-Changed World." In *Curating the Future: Museums, Communities and Climate Change*, edited by Jennifer Newell, Libby Robin and Kirsten Wehner, 1–16. London: Routledge.

Paaskoski, Leena. 2020. "Dynaaminen museo kartuttaa tulevaisuusperintöä." *MuseoPro*. Accessed 6 December 2023. https://www.museopro.fi/fi/dynaaminen_museo_kartuttaa_tulevaisuusperintoa

Paaskoski, Leena, Katriina Siivonen, Noora Vähäkari, Pauliina Latvala-Harvilahti, Päivi Pelli, Maria Granlund and Teppo Hujala. 2025. *The Dynamic Museum and Heritage Futures Workshop. A Handbook for Eco-social Bildung Work in Museums*. Publications of Lusto 9. Savonlinna: Finnish Forest Museum Lusto.

Patokorpi, Erkki and Marko Ahvenainen. 2009. "Developing an Abduction-based Method for Futures Research." *Futures* 49: 126–139.

Peirce, Charles Sanders. 1994/1931–1958. *The electronic reproduction of The Collected papers of Charles Sanders Peirce*, edited by John Deely including earlier publications: The Collected papers of Charles Sanders Peirce, Vols. I–VI, edited by Charles Hartshorne and Paul Weiss, Cambridge, MA: Harvard University press, 1931–1935 and The Collected papers of Charles Sanders Peirce, Vols. VII–VIII, edited by Arthur W. Burks, Cambridge, MA: Harvard University Press, 1958.

Poli, Robert. 2011. "Ethics and Futures Studies." In *International Journal of Management Concepts and Philosophy* 5/4: 403–410.

Poli, Robert. 2017. *Introduction to Anticipation Studies*. Cham: Springer.

Richardson, Katherine, Will Steffen, Wolfgang Lucht et al. 2023. "Earth beyond six of nine Planetary Boundaries." *Science Advances* 9. https://www.science.org/doi/10.1126/sciadv.adh2458

RMIT University. 2017. *The triple bottom line or the three P's*. Accessed 17 October 2024. https://www.futurelearn.com/info/courses/sustainable-business/0/steps/78339#:~:text=The%203Ps%20of%20sustainability%20are,programs%20take%20care%20of%20people

Rockström, Johan, Will Steffen, Kevin Noone et al. 2009. "A Safe Operating Space for Humanity." *Nature* 461: 472–475. https://doi.org/10.1038/461472a

Rubin, Anita. 2000. *Growing up in Social Transition: In Search of a Late-modern Identity*. Annales Universitatis Turkuensis, Ser. B, Tom. 234, Humaniora. Turku: University of Turku.

Siivonen, Katriina. 1994. "Budskapet i människors yttre. Ett tillämpningsförsök av semiotisk analysmetod." *Nord Nytt* 55: 86–95.

Siivonen, Katriina. 2008a. *Saaristoidentiteetit merkkien virtoina. Varsinaissuomalainen arki ja aluekehitystyö globalisaation murroksessa.* Helsinki: Suomen muinaismuistoyhdistys.

Siivonen, Katriina. 2008b. "Culture is Basically Creative." In *Creative Futures Conference Proceedings. 10.–11. October 2007 in Pori, Finland*, edited by Marjo Mäenpää and Taina Rajanti, 180–190. Taideteollisen korkeakoulun julkaisu C 6. Pori: University of Art and Design, Pori School of Art and Media.

Siivonen, Katriina. 2017. "Kulttuurinen kestävyys." In *Kaikki vapaudesta*, edited by Kari Enqvist, Ilari Hetemäki and Teija Tiilikainen, 275–287. Helsinki: Gaudeamus.

Siivonen, Katriina. 2018. "Sustainable Everyday Culture from Glocal Archipelago Culture." In *Cultural Sustainability and the Nature-Culture interface. Livelihoods, Policies, and Methodologies*, edited by Inger Birkeland, Rob Burton, Constanza Parra and Katriina Siivonen, 19–34. London: Routledge.

Siivonen, Katriina. 2019. "Intangible Cultural Heritage will Revolutionise the Future." In *Wohin geht die Reise? Where does the journey go to? Eine Geburtstagsgabe für Johanna Rolshoven 2019, A birthday present for Johanna Rolshoven*, edited by Sabine Eggmann, Susanna Kolbe and Justin Winkler, 93–97. Basel: Akroama Editions. https://www.geruchderzeit.org/siivonen/

Siivonen, Katriina. 2021. "Cultural transformation and cultural sustainability." Lecture held at University of Turku, Turku School of Economics, Master's Degree programme in Futures Studies, as a part of the course *FUTUS2 Theories of Futures Studies and Transformation* on 30th September 2021, and on fall terms 2022, 2023 and 2024.

Siivonen, Katriina. 2022. "Tulevaisuusperintö kulttuurisen kestävyysmurroksen välineenä." In *Tulevaisuudentutkimus tutuksi – perusteita ja menetelmiä*, edited by Hanna-Kaisa Aalto, Katariina Heikkilä, Pasi Keski-Pukkila, Maija Mäki and Markus Pöllänen, 419–433. TVA-julkaisuja 1/2022. Turku: Tulevaisuudentutkimuksen Verkostoakatemia. https://urn.fi/URN:ISBN:978-952-249-563-1

Siivonen, Katriina, Pauliina Latvala-Harvilahti, Noora Vähäkari, Leena Paaskoski and Päivi Pelli. 2022. "Tulevaisuusperintöverstas kestävän tulevaisuuden edistäjänä." In *Tulevaisuudentutkimus tutuksi – Perusteita ja menetelmiä*, edited by Hanna-Kaisa Aalto, Katariina Heikkilä, Pasi Keski-Pukkila, Maija Mäki and Markus Pöllänen, 250–259. Tulevaisuudentutkimuksen Verkostoakatemian julkaisuja 1/2022. Turku: Tulevaisuudentutkimuksen Verkostoakatemia. https://urn.fi/URN:ISBN:978-952-249-563-1

Siivonen, Katriina, Hannele Cantell, Outi Ihanainen-Rokio et al. 2024. "The Means of Achieving Sustainability Transformation." In *The Avenues to Sustainability. Key Lessons from Sustainability Science,* edited by Tarja Halonen, Kaisa Korhonen-Kurki, Jari Niemelä and Janna Pietikäinen, 215–254. Translated by Suvi Lintuvaara. Helsinki: Gaudeamus.

Simmel, Georg. 1988. *La tragédie de la culture et autres essais.* Paris: Rivage.

Smith, Laurajane. 2006. *Uses of Heritage.* Abingdon: Routledge.

Steffen, Will, Katherine Richardson, Johan Rockström et al. 2015. "Planetary Boundaries: Guiding Human Development on a Changing Planet." *Science* 347: 6223. https://doi.org/10.1126/science.1259855

SUSTEMY. 2024. *3 Principles of Sustainability: Building a Greener Future.* Sustainability Education Academy. Accessed 17 October 2024. sustainability educationacademy.com/3-principles-of-sustainability-building-a-greener-future/#

Thomas, Suzie, Vesa-Pekka Herva, Oula Seitsonen and Eerika Koskinen-Koivisto. 2019. "Dark Heritage." In *Encyclopedia of Global Archaeology*, edited by Claire Smith. Cham: Springer. https://doi.org/10.1007/978-3-319-51726-1_3197-1

UNESCO. 1972. *Convention Concerning the Protection of the World Cultural and Natural Heritage*. https://whc.unesco.org/en/conventiontext/

UNESCO. 2003. *Convention for the Safeguarding of the Intangible Cultural Heritage*. https://ich.unesco.org/en/convention

UNESCO. 2014. *Shaping the Future We Want. UN Decade of Education for Sustainable Development (2005–2014). Final Report*. Paris: UNESCO.

United Nations. 1948. *Universal Declaration of Human Rights*. https://www.un.org/en/about-us/universal-declaration-of-human-rights

United Nations. 1987. *Report of the World Commission on Environment and Development: Our Common Future*. New York: United Nations.

United Nations. 2023. *Sustainable Development Goals*. https://sdgs.un.org/goals

United Nations. 2024. *Sustainability*. https://www.un.org/en/academic-impact/sustainability

White, Rehema. 2013. "Sustainability Research: A Novel Mode of Knowledge Generation to Explore Alternative Ways for People and Planet." In *The Sustainable University: Progress and prospects*, edited by Stephen Sterling, Larch Maxey and Heather Luna, 194–217. London: Routledge. https://doi.org/10.4324/9780203101780-11

White, Rehema, Ullrich Kockel, Betsy King, Kirsten Leask, Peter Higgins and Andrew Samuel. 2023. "The Future Starts in the Past: Embedding Learning for Sustainability through Culture and Community in Scotland." *Frontiers of Sustainability* 4. https://doi.org/10.3389/frsus.2023.1128620

Wiktor-Mach, Dobrosława. 2018. "What Role for Culture in the age of Sustainable Development? UNESCO's Advocacy in the 2030 Agenda Negotiations." *International Journal of Cultural Policy* 26 (3): 312–327. https://doi.org/10.1080/10286632.2018.1534841

Willamo, Risto, Leena Helenius, Charlotta Holmström, Liisa Haapanen, Essi Huotari and Vilma Sandström. 2017. "Kuinka ymmärtää kompleksisia ilmiöitä? Generalismi, holismi ja holarkismi kokonaisvaltaisessa kestävyyskoulutuksessa." *Kasvatus* 48/5: 415–428.

3 Ecosophy, Geopoetics, and the Desire to Know

Mary Modeen

What constitutes the various aspects of knowledge of one's place in the world – the land, environment, co-participation with other-than-human beings and ability to understand the interconnectedness with one's fellow cohabitants – is the focus of this chapter. In reaching to include more than the sensory, embodied experience of the phenomenal world, the reflection here moves across ontologies of intuition, metaphysics, various forms of indigenous and traditional knowledge (TK), and spiritual reflection, all of which are important inclusions. Qualities of being that lie in the realm of ineffable wisdom are touched upon with examples from art and poetry that point to these wordless comprehensions in creative responses to knowingness. Aspects of the earth itself, as couched in philosophical modes of knowing, such as geopoetics, ecosophy, and the emergence of Whole Earth Systems science are also included. Weaving in between these musings, and between the visible and invisible world we inhabit, Nan Shepherd's words in *The Living Mountain* resound like chiming bells, returning us again and again to the attentiveness of being-in-place.

The Living Mountain

To know 'the essential nature' of her beloved Cairngorm Mountains (see Figure 3.1), Nan Shepherd celebrated living her life fully in *The Living Mountain*.[1] She specified that she set out walking in the mountains, '[t]o know, that is, with a knowledge that is a process of living' (2011, 3). In other words, the knowledge that she embraced was resolutely shaped by lived experience. The manner by which she aimed to achieve her investigation was through embodied 'living', which she specifies with these words (Shepherd 2011, 3):

> This is not easily done in an hour. It is a tale too slow for the impatience of our age, not of immediate enough import for its desperate problems. Yet it has its own rare value [...] However often I walk on them, these hills hold astonishment for me. There is no getting accustomed to them.

DOI: 10.4324/9781003727972-3

Figure 3.1 View of Braeriach, Cairngorms.

Source: Photograph by Andy Rice (2007), used by permission.

Thus, in the first paragraph of her remarkable book, Shepherd set out her aim to derive knowledge that is based on first-hand experience, necessitated by and through embodied life itself, and in ways that were equally physical and metaphysical, with empirical discovery as well as through aesthetic and emotive experience ('astonishment'). In doing so, she eschewed habit, seeking to see and know afresh with each journey.

Shepherd's text is key to this chapter because it crystallises the many qualities of knowledge itself, which merges embodied, observational experience with aesthetic, emotive, intuitive, and speculative knowledge. This synthesis has found expression in many forms, and examples which follow will draw out these human-crafted representations of complex and multifaceted ways of knowing through ecological, poetic, and artistic modes of representation. At its core, the focus is also on the earth itself, with its myriad life forms and geologic base shifting over time. These elements are profoundly interconnected and demand a layered perspective and mixed approach of cultures, languages, and disciplinary orientations to explore the complexity that Shepherd implied in her term 'Living', since it is both a research focus *and* the quality she ascribes to the mountains.

A word is in order here about the terms in this chapter that may seem very familiar, but which are not interchangeable among themselves. The 'desire to know' is the innate curiosity that drives people to understand their surroundings, to investigate and 'research' in depth (whether academically or empirically), using many techniques and modes of inquiry, and then to act in response to that 'knowledge' and knowingness. Forms of knowledge, as indicated above, may themselves take many shapes, verbal and non-verbal, performed and created. In this chapter, the aim is to typify and celebrate the kindred spirit of the 'living' quality Shepherd so ardently embraced, honouring the many forms in which they appear in the multi-modal cross-disciplinary examples which are offered herein. A final term to clarify is the pronoun 'we', used here despite the single authorship of this chapter. This author uses 'we' in the wish that readers will consider their own experiences in the appreciation of the philosophical and socio-political consequences suggested in this text. The 'we' pronoun should be considered 'poetic', in offering an invitation for engagement, appealing to your imagination, including you, the reader, to participate. But it is not a disingenuous inclusion, and never assumes your suspension of critical reading.

Before leaving this introduction, Shepherd (2011, 4) reminds us that:

> One cannot know the rivers till one has seen them at their sources; but this journey to the sources is not to be taken lightly. One walks among elementals. And elementals are not governable. There are awakened also in oneself by the contact elementals that are as unpredictable as wind or snow.

Her caution that is specific to our intention here is to acknowledge that 'the elementals are not governable', that it is not in humankind's ability to recast the interconnections of the entirety of the seen and unseen ecosystem that applies to Shepherd's notion of 'a living landscape'. She also tells us plainly that investigating this knowledge, which is so demanding and extensive in its pursuance, will inevitably change us as knowledge-seekers as well. Although in her Introduction she does not set out how this change occurs, the cumulative experiences she shares throughout the book lead the reader to understand that these changes are as much about self-knowledge as they are about geographical or aesthetic knowledge, and perhaps a tacit acceptance of transcendent knowledge as well.

Let us momentarily note some of the separate strands in this tapestry of 'living knowledge' we seek to investigate. The first thread among philosophical treatises to follow came to prevalence in the seventeenth and eighteenth centuries and concerns the epistemology that relies upon observational information gained in empirical methods. Direct observation and subsequent inductive reasoning, based on this information, most typically are the basis of scientific investigation, and are fundamental to research in the natural and social sciences.

A second, closely related thread we might trace is that of phenomenology, as developed in the twentieth century, emerging from the first works by Husserl. This may be described as a philosophy of essences, as 'phenomenology is also a philosophy which puts essences back into existence' (Merleau-Ponty 1962, Preface). Merleau-Ponty continues to describe this as a 'transcendental philosophy', 'a philosophy for which the world is always "already there" before reflection begins – as "an inalienable presence"' (1962, Preface). It is based on a 'lived' experience and, as such, Merleau-Ponty's promotion of the embodied and first-hand experience not only presages Shepherd's words, but also forges the links between unique sensory experiences, the human 'felt' experience, and a practice of phenomenology. This combines the realisation that the perceiver simultaneously perceives the environment while being aware of *doing* the perceiving, and is participant as a part of the total perception. As such, the observer affects the conditions of that which is being perceived and is also changed by the act – which Shepherd also clearly appreciates.

In these philosophical variations, bodily senses play a key part in the ways in which the world is taken in, sensed, and perceived, enabling a consciousness or awareness of the environmental phenomena. Sight, sound, smell, and touch combine in the perceiver, facilitating sense making, identification, experiential memory, expectation, and emotive and intuitive felt responses. Shepherd, for example, speaking of her feeling of *feyness* in mountain climbing, said (Shepherd 2011, 6–7):

> At first, I had a thought that this lightness of body was a universal reaction to rarer air. It surprised me to discover that that some people suffered malaise at altitudes that released me, but were happy in low valleys where I felt extinguished. Then I began to see that our devotions have more to do with our physiological peculiarities than we admit. I am a mountain lover because my body is at its best in the rarer air of the heights and communicates its elation to the mind.

Her term 'feyness' has a particular significance. Partly, she confesses, it is being 'a little mad', carried away with 'the joyous release of body'. But other connotations of the word 'fey' are 'transcendence', 'otherworldliness', and 'ethereal'. The paradox of being most keenly *in* the world while feeling otherworldly is not lost on a reader. In Scottish parlance, 'fey' can also mean 'fated', 'hapless', or 'doomed'. It is in this association that Shepherd senses her haplessness, which she intuits may have been the mysterious effect of something ineffable that moved between her and the mountain (Shepherd 2011, 7; my italics):

> [T]here is more in the lust for a mountain top than a perfect physiological adjustment. What more there is lies within the mountain. Something moves between me and it. *Place and a mind may interpenetrate till the nature of both is altered.*

The mountain's rarefied air, the vastness of the space, and her own bodily lightness affect her deeply in this example of a co-constructed, co-created, multifaceted knowledge of her 'Living Mountain'. But more than that is her open-ended perception that the mountain has been affected as well, in the exchanged 'something' between mountain and walker.

Walking has long been regarded as a means of investigation and knowing, together and separately, in part through the power of that which is encountered to stimulate thought, but equally through the body's stimulation as well. Nietzsche, who was a keen walker, was quoted as saying, 'Give no credence to any thought that was not born outdoors while one moved about freely [...] in which the muscles are not celebrating a feast too' (Nietzsche 1969, 239–240). The walking encounter with topography and the environment stimulates the mind and allows one to receive the benefits from fresh air and moving the body. Walking philosophers are a well-known phenomenon, and Bruce Baugh's book, *Philosopher's Walks* (2022), reviews a good many such in the course of modern history, and the significance of walking to their works. Baugh attests: 'More than just perception of what is present to the senses, then, walking is a remembrance of the walks of precursors who first laid out these paths' (Baugh 2022, 8). He makes clear that 'precursors' are more-than-human as well as human, listing deer, coyote, bear, cougar, and moose. I would include among these walking enthusiasts further writers who link this peripatetic activity to forms of knowledge: Rebecca Solnit, Dee Heddon, David Borthwick, and Mike Collier, among others.

Artists such as Richard Long and Hamish Fulton, and more recently, Judith Tucker, Harriet Tarlo, Rebecca Thomas, Gareth M. Jones, and Thomas A. Clark, among many, represent a multitude of walking artists and poets who also embrace walking as both a primary research method, immersed in changing environments with the pace of their peripatetic movements, and the means of knowing. By 'research' in this case, I refer to the encounter that is intrinsic to various types of contemporary walking practices. Encounter can be paced with bodily exertion, timed with breaths and strides, and present the walker with the entirely unexpected; a chance ray of sun breaking through the clouds, a kestrel fluttering overhead, or culturally distinctive objects such as *kekkai,* Japanese border markers, for example. Nor are the experiences embedded within the encounter constrained to humans alone. Artist/academic Judith Tucker (2020, 142) wrote:

> [H]ow might such an embodied methodology [walking] which in some ways privileges the (able-bodied) human body, remain open to considerations of the post-human or the more-than-human, including inter- or trans-species communication?

Her answer to this is framed thusly, in speaking of her collaborative walking practice with poet Harriet Tarlo; she explains that she considers their work 'not only in the context of critical contemporary walking practices, but also within the context of the ecocritical turn, via the phenomenological, towards New Materialist thinking' (Tucker 2020, 142). Drawing on the work of Haraway, Barad, 'and on Stacy Alaimo's trans-corporeality, they develop an idea of trans-materiality of which does not place the human at the centre' (2020, 142). Tucker's paintings are informed by the depth of a reflective walking practice, in her case capturing the sights and species in the Yorkshire moors, while delving ever more intensively in the desire to know more.

Tucker's art demonstrated her ability to hint at different traces of activity that can be studied in her paintings, but which are not always immediately apparent or 'remarkable', such as upturned earth or pollarded trees, or lights switched on in distant houses. They reveal through absence as much as by what is represented, but always foreground observation.

For the artist, the walker, and the keen observer, the ways that phenomena are encountered become the materials for contemplation and interpretation, thereby leading to that which is framed in one's understanding. This framing of what is encountered fuels thinking that follows, and constitutes the construction of the experience, the 'nature' and environment upon which it is based. Contemporary critical walking practices then, facilitate and promote experience based on first-hand encounter, valuing participatory experiential knowledge, in the manner of Shepherd.

To share these perceptions, and perhaps to speculate on Shepherd's 'something' that has been exchanged between mountain and walker, an explosion of new nature writing is underway across the world; in the UK, in particular, this growing body of literature combines many of the similar concerns which Shepherd foregrounds, and which have been taken up by a good many other writers.[2] Performative actions by creative artists and eco-activists, and sited art interventions in the landscape, have blossomed in recent years (Modeen and Biggs 2020). Creative works in various media and genres allow for the incorporation of inspiration, materiality, and actions to address this complexity of types of knowledge. Moving through the landscape, the plurality of modes in sensory, metaphysical, emotive, and affective states form an intersection of cognitive awareness that is the foundation of being-in-the-world. This capacity for human knowledge is not distinct from the environment; rather, it is inherent in being-in-the-world, a part of it, while 'seeking to know', as Shepherd says, 'its essential nature'. And to this end, she emphasised that it is 'a grand thing to get leave to live' although at times, like climbing a mountain, not always easy, 'it is tough bitter going'.

Shepherd never moves far from the embodied perspective, or the recognition of 'tough, bitter going'. When she says, 'our devotions have more to do with our physiological peculiarities than we admit', what she also implies is the specificity of each individual. There are some who cannot breathe or feel exhilarated in higher altitudes, and some who (like her), clearly do. Perception is for those who are able to perceive. Mountaineering demands ableist abilities, or at the very least, extraordinary support for those who might avail themselves of assistance to ascend the heights.

One of the writers who has emerged in the explosion of nature writing has been Robert Macfarlane. He has been deeply influenced by Shepherd's writing and cites two of his favourite ideas that she proposed: the first is that there is an 'inside' to mountains, and that one does not walk up but *into* a mountain. The second is that one needs to abandon the notion of a summit as a goal and instead to be a 'peerer into the nooks and crannies', in the totality of the experience. 'Sanctuary, solitude, and a relationship with place', as Macfarlane discusses on a BBC radio broadcast, stand as Shepherd's beacons of understanding and experience (Macfarlane 2013b). Macfarlane was so taken with *The Living Mountain* that he was instrumental in its reprint (Shepherd 2011), and wrote the new Introduction, along with Jeanette Winterson, who contributed an Afterword. It is also notable that Samantha Walton has just published a book on the same text (Walton 2021). In this re-visitation of Shepherd's ideas, Walton expands the context from Shepherd's post-War Scotland and nascent environmental schemes to touch on contemporary issues through eco-critical approaches to Deep Time, vital matter, and selfhood. Walton concurs with much of this author's assessment, adding that 'Shepherd can be seen as a literary phenomenologist, who developed her own nuanced understanding of mind-body relations and embodied subjectivity [...] and accounts of embodied experience' (Walton 2021, 172).

The fact that we humans are 'always already' in the world before we come to know this, or to know ourselves, lies in part as the curiosity behind the driving forces of art making, reading philosophy, and searching for meaning. As living organisms, we ponder this dual nature of being both subject and object, of being the perceiver and the perceived, while seeking to know the nature of being alive in the environment we inhabit, whether climbing a mountain, negotiating a body of water, or wandering city streets. Merleau-Ponty (2002, 37) remarks:

> The *real* is a closely woven fabric. It does not await our judgement before incorporating the most surprising phenomena [...] Perception is not a science of the world, it is not even an act, a deliberate taking up of a position; it is the background from which all acts stand out, and is presupposed by them.

Our focus on embodied perception may overlook important information if we forget to pay attention to the invisible, the silent, or the odourless; in short, the world that is extra-bodily, beyond unaided human detection. In the next section, we will delve into the imperceptible aspects of the environment, at least to those phenomena that require the aid of microscopes and other more-than-human sensing detection capacity.

Geopoetics and Invisible Interactions

A working definition of geopoetics would be useful here: 'geopoetics' is a movement to comprehend being through intellectual, sensory, and intuitive understanding, crossing cultures, species, and geographies, seeking to communicate in more than language. With the understanding that there are ways of knowing that transcend language, works of art, music, dance, and performance reach with a deep ability to speak to these other ways of knowing. Equally, embracing geopoetics means acknowledging that the earth is, in its own multifaceted reality, so nuanced, complex, and changeable that its character might be unknowable in its whole-earth sense, recalling Shepherd's words that 'elementals are ungovernable'.

Living in an enmeshed world means being inextricably connected to all beings and things. Existing as a sentient, cognitively attuned individual, one can see, hear, smell, touch, and even taste the abundant complexity in the immediate environment. It is what Shepherd refers to as 'sensational' and 'perpetually new' in describing her experiences in the Cairngorms. This interconnectedness, a term used consistently here throughout this chapter, is also linked with *ecosophy*, in the sense that Félix Guattari coined, meaning that it is necessary to consider humans in relationship with their environment (Guattari 2000). In Guattari's use of the term, the parallels between ecological environments were entangled with social environments and mental environments, setting out a case for enmeshed, inter-relational states of being. In the complex inter-relationships between an individual and its existence in its environmental context, there can be no real separation between a single entity and its surroundings. The term 'ecosophy' was coined by both Félix Guattari and Arne Naess, separately, as a contraction of ecological philosophy. As a neologism, its intent was to highlight the need for combined philosophical and environmental wisdom. It is important to note that environmental elements go far beyond what humans can detect with the body alone. For example, one aspect of the natural environment that is often invisible to the naked human eye is the complex, richly diverse nature of mycelial entanglement.

Merlin Sheldrake, in his book *Entangled Life*, lays out the wonders, mysteries, and overall omnipresence of this life kingdom of species (2020, 3). As a mycologist, his desire parallels Shepherd's in passion, dedication, and the desire to know the essential nature of fungi. His language at times is very similar to Shepherd's as well, conveying a sense of wonder, deep fascination, and appreciation for fungi. He begins (Sheldrake 2020, 3):

> Fungi are everywhere but they are easy to miss. They are inside you and around you. They sustain you and all that you depend on [...] Fungi provide a key to understanding the planet on which we live, and the ways we think, feel and behave. Yet they live their lives largely hidden from view.

He proceeds to explain in the introduction the broadest possible terms of entanglement and connectivity, among and between fungal forms, and details various examples of physical structures and shapes that facilitate web-like phenomena. Another task Sheldrake undertakes is that of challenging attitudes based on limited human understanding of fungi, preferring to move away from associations with death or decay, and to promote an entirely more positive understanding of being 'looked after' by other fungi and other species. Since the importance of the various species in the fungal world is essential to all ecosystems, Sheldrake insists that we need to attend to the environment to defend our 'protectors and guardians'. He wryly calls the intricate inter-relationships that exist between mycelial forms and the other life kingdoms the 'wood wide web' (2020, 4). This attention to the minuscule, and its importance in unseen but omnipresent microbial life, is the foundation for the film entitled *Our Body is a Planet*, made by Léonie Hampton (2022). The notion of an indivisible world is paramount in this work, looking at many ways in which human bodies are hosts for countless organisms, and as such, comprise a life form that is multiple and complex. Understanding the diversity of DNA differentiation, she recites the phrase that 'overwhelmingly, our bodies are more microbial than human' (2022, 04:43–04:47).

In the word 'ecosystems', the concept of a whole environment with every component element was emphasised. The word was first used in 1935 in a publication by British ecologist Arthur Tansley; the term was coined by Arthur Roy Clapham, who devised the word at Tansley's request (Willis 1997). It was used to underscore the interactions of any single aspect within an environment with its co-participant neighbours and conditions. Tansley wished it to be understood that any characteristics of a particular plant or animal element within the whole could only be known by its relational qualities with all the inorganic parts and conditions, including conditions of heat and wind. Of course, this work has underpinned ecological enquiry since Tansley's publication, but it is not new to human knowledge. It is significant that Indigenous knowledge systems for hundreds, if not thousands of years have featured the interconnectedness of all things, human and non-human, in a living earth (echoing Shepherd's 'living mountain'). More on this below. Similarly, in TK and local expert knowledge (LEK) the practical actions of farmers, fishermen, and hunter-gatherers, for example, have depended for millennia on precisely understanding, predicting, and adapting to these changeable conditions.

Pursuing 'the desire to know' with scientific research methods embraces a distinctive type of knowledge; while it excludes belief systems, emotive responses, and the language of poets it has its own defined field that most closely studies vastly complex interactions on the planet. In 1965, James Lovelock began to develop the Gaia theory, and continued in the late 1960s and early 1970s with the collaboration of Lynn Margulis, a microbiologist. In reference to the Gaia hypothesis, Tim Lenton (2016, 5) explains that '[i]t represents the first scientific statement of the Earth as a system that is more than the sum of its parts. Thus for me at least, the Gaia hypothesis marks the start of Earth system science'. And in Hamer Dodd's artwork (see Figure 3.2), the artistic vision of an unseen but enormously complex and patterned 'world with the world' exists in every substance.

Lenton (2016, 5; my italics) adds in the debate surrounding the Gaia theory:

> Although they (Lovelock and Margulis) didn't intend it, this (hypothesis) seemed to imply a sort of purposive control of the global environment by unconscious organisms. Such teleological reasoning is *out of bounds in science.*

Figure 3.2 Hamer Dodds, ***Structural Intuitions, No. 1***, ink on paper.

Source: Artwork by Hamer Dodds (2022). As an interdisciplinary artist/biologist, Dodds' image here is a speculative visual response to the entanglement of life forms and structures, intuited in speculative imagination.

It is a field now identified as Earth Systems Science. It may be said to approach the study of all of the interactions on Earth as holistically implicated, including life forms, geological elements, chemical interactions, and the effects from the sun. Lenton's concise introduction to the enmeshed qualities in Earth Systems Science also features qualities that life brings to the planet, tracing cycles of life and decay across and in planetary changes over time (Lenton 2016). Therefore, time is an additional factor necessary in understanding Earth Systems, particularly when considering geological epochs or plate tectonics.

This emphasis on entanglement and enmeshed life forms links many of the specific forms of knowledge. Interconnectedness, as suggested above, has been at the core of most world Indigenous belief systems long before records began. In oral traditions, for example, the long told and oft-repeated stories of creation and first ancestors play out many variations of interconnectedness between humans, non-humans, and animated organic matter in our environment. Kimmerer (2013b) writes: 'The trees act not as individuals, but somehow as a collective. Exactly how they do this, we don't know. But what we see is the power of unity. What happens to one happens to us all'.

Traditional Indigenous and LEK keepers are unified in their belief systems that stress interconnectedness.[3] In these traditional beliefs, and shared through storytelling and art, humans are innately interconnected with non-humans, with their land, with water, with air, with the sun and the moon, with communities of others, and in understanding this, they accept their position as stewards. 'Everything has a spirit', say the elders, 'and is the life force of every living thing'. This is not to suggest that all Indigenous belief is identical; in Canada, for example, distinctions are drawn between First Nations people's beliefs, Inuit, and Métis, with nuances of values, attentiveness, and priorities. But each distinctive culture echoes the message of intra- and inter-relations between humans, animals, and all things.

These 'spirits' of Indigenous belief that imbue all living things have a partial correspondence energy as studied in quantum physics, and in the science of mycology. While we cannot always see the life forces and cycles of fungal growth with the unaided human eye, for example, the complexity and vigour of mycological species is awe-inspiring.

Coming from the perspective of an arborist and mycologist, Wohlleben's concept of both interconnectedness and communication across trees, plants, and fungi (echoing Kimmerer's words above), was also repeated in recent works, with forestry research deployed to establish and explain the power of communication between and among trees and animals (Wohlleben 2017; also, Wohlleben 2019).

Another important contemporary work which merges visualisation and mycology research, Hampton's film, *Our Body is a Planet*, reminds the viewer that it was the microscope that allowed the view of the microbial world (Hampton 2021; also, see Figure 3.3). But before we discuss this video and current medicinal mycology, it is worth noting that it was

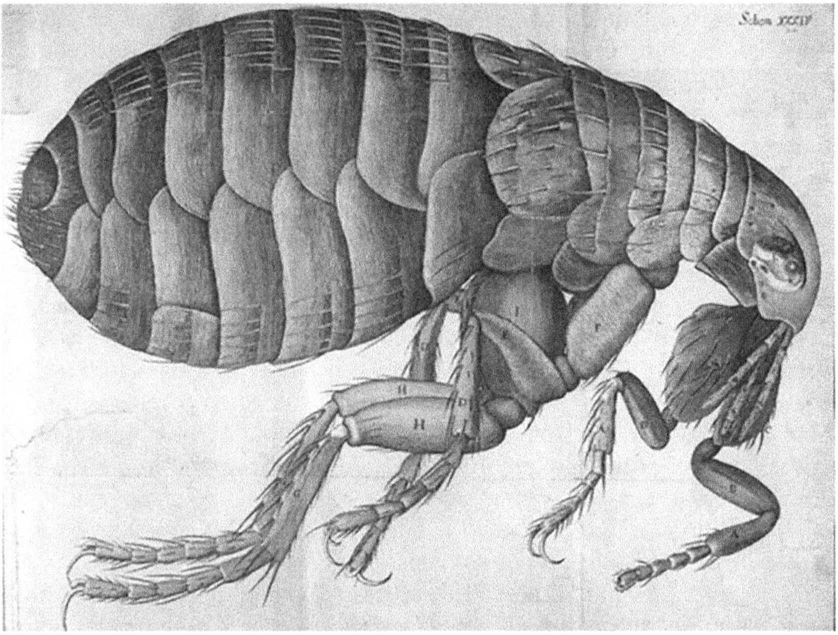

Figure 3.3 Hooke, R. *Flea*, from *Micrographia* (1665).
Source: Courtesy of the British Library.

Antonie Philips van Leeuwenhoek (1632–1723) who created the lenses which facilitated the viewing of microbes – what he called 'the flowers between the teeth' which he was certain would frighten an ordinary person. Using this invention, it was Robert Hooke (1635–1703) in his book *Micrographia* (meaning literally, 'small pictures'), who coined the term 'cells'. To be precise, the entire title of his publication (Hooke 1665) is *Micrographia: Or Some Physiological Descriptions of Minute Bodies Made by Magnifying Glasses With Observations and Inquiries Thereupon.* The illustrations include painstakingly detailed drawings of a flea, a louse, and the cellular structure of cork, among other things, which he drew from his magnified observations.

The reference to 'cells' was used because the living organisms he was viewing reminded him of monks in their solitary rooms, also called cells, within the monastery. In fact, the 'cell' as he termed it, is still taught as the building block of all life. Now, over 350 years after van Leeuwenhoek and Hooke, Hampton's film, which visually displays images produced with the technological enhancement of electronic microscopy, also lays out the pervasive scale of cellular life, citing 'a gram of cells has more life forms than all the people on earth'. From small to smaller to the most minute living forms, the desire to know has driven the development of technologies by which to learn even more about the entanglements of life forms.

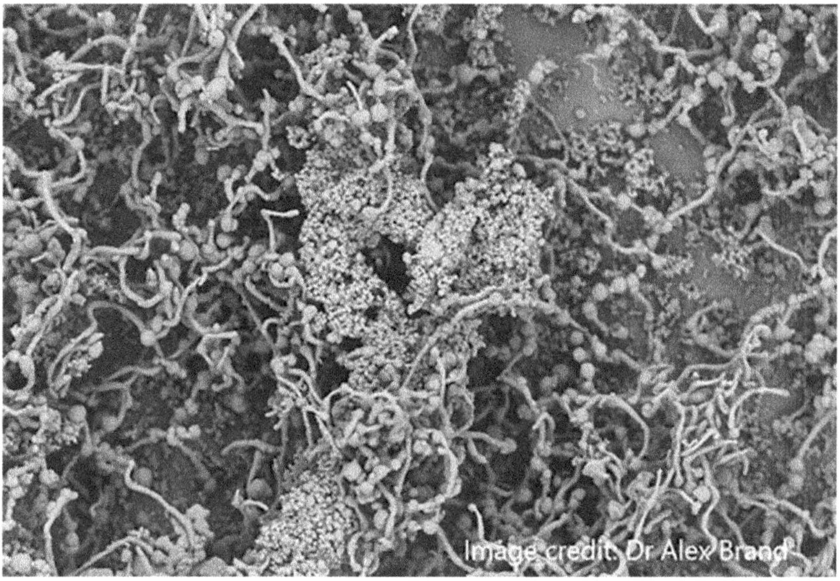

Figure 3.4 A Mycelial mesh, taken with an electron microscope.

Source: Photograph by Alex Brand. The organisms are a mix of the commensal fungus, *Candida albicans* (which forms the long filaments called 'hyphae') and various strains of bacteria that can cause infections in humans.

The struggle of living organisms to survive is omnipresent, but not all struggles are equally effective. Fungi are supremely successful in their mycelial symbiosis within their environment (Hampton 2022, 06:06–06:16): 'Modern humankind think we can adapt to challenging situations by changing our environment, whereas fungi change themselves'. Given this ability to survive, adapt, and respond, exploring space through growth, and communicating through chemical changes, it might be considered that fungi are the most successful of *all* life forms, and almost 'immortal' as we understand that term, in its ability to grow whole networks from a single cell. In the microscopic image (see Figure 3.4) the enmeshed network of mycelial growth is evident.

The capacity to form symbiotic partnerships with other living species is a way for fungi to collaborate in resilience and survival. But it also makes them vulnerable, to a degree, when conditions change suddenly. We have all seen how precipitous change is becoming more frequent in our world, with climate changes, radical weather shifts, species' extinctions through loss of habitats, and contamination through the overuse of fungal pesticides and other pollution. If these are inter-related effects in a complexly interdependent series of ecosystems – whether that is one's own body as a host to millions of organisms, or in the external natural environment where changes are rapid and extreme – then the possibilities of unexpected mutations and pathogens can increase significantly. 'Fungi also brought us penicillin, reminding us that it can cure as well as kill. The sense of equilibrium between toxin and remedy is key to our survival' (Hampton 2022, 08:54–09:10).

In focusing on the interconnected and relational aspects of ecosystems, on both the micro- and macro-environments, one experiences a sense of awe at the complexity of interactions between species, inorganic matter, and energies. This understanding of co-relational interactions has led to various contemporary paradigms of knowledge structured alternately as 'deep systems', 'Gaia', 'Earth System Science', 'ecosophy', and 'geopoetics', among others.[4] These, and many other models of conceptualising complex co-relational interactions inherent in states of being, have emerged as an increasingly urgent desire to find new means of rebalancing sustainable modes of life on this planet. Hampton (2022, 09:17–09:23), echoing the concern of millions, asks: 'Can we learn from other species to heal and rebalance our world?'

Experience, Singular and Collective, and the Desire to Know

This question of learning and the desire to know is where we began this chapter, citing Shepherd's passion to know the Cairngorms as a 'living' experience, to know 'with a knowledge that is a process of living'. Shepherd embraces how we humans come to know what we know as an active engagement, a co-participatory experience of being-in-the-world as a fully sentient co-being (or 'co-actant', as some would express it). What is useful in tracing this desire to know, affirming sustainable ontologies, are the distinctions to be made by the individual as singular ('my mountain' or 'my place') as opposed to the collective ('our environment' and 'our actions'). It may be said that we are both, as contradictory as that might initially seem.

The individual experience informs the collective. This we know to be true, because since infancy, we have heard stories, and later read and watched, narratives of personal experiences unfold. Countless writers have told of their experiences in the natural environment, documentary filmmakers have recorded the world in dramatic imagery. But we have also been party to the aggregation of experiences that allow us to move from the singular and particular to form a community of shared experiences. It is the power of human imagination to make this leap, to learn from others, and to value the universal truth embedded within the unique, retold experience. As James Cox implies, we are generally guided by 'the phenomenological mantra that nothing human ultimately is alien to other humans' (Cox 2018, xix). In other words, to be human is to recognise what is comprehensible in the experience of other humans. Cox's writing focuses on the ethnographic work of T.G.H. Strehlow, who lived and researched with the Elders of the Indigenous Arrernte people of Central Australia in the first half of the twentieth century. As Cox explains, Strehlow was born to missionary parents who were based in Central Australia, and grew up speaking English, German and the local Western Arrernte language. Although he was non-Indigenous by birth, Strehlow's status was more that of a non-native 'insider', as he studied

Arrernte language, religion, artefacts, and site-specific belief systems, that included complex kinships with the land and non-humans. Although criticised by some for transgressions by sharing secret-sacred knowledge with an outside world, Cox weaves together the legacy of Strehlow's work to save the Arrernte's complex belief systems, with collected objects and ceremonies, linking people with their geographically specific natural environment, connecting a 'personal monototemism in a polytotemic community'. This personal identity and lifelong totemic alignment were determined through the exact location of 'where the future mother first became aware that she was pregnant' (Cox 2018, 60). In this one example of Arrernte traditional belief, the place of one's conception is inextricably linked to a totemic identity and personal responsibility to the natural world, to one's own social community, and to an orientation for a spiritual alignment. Through ceremonies and daily practices, this enmeshed quality of being-in-the-world becomes real for its practitioners, endowing them with the phenomenologically embodied personification of a living state and 'time outside of time', simultaneously now in the moment and in time 'out of all eternity'.

Time is intrinsically linked to being alive in each sited place. The paradox of 'timeless time' in comparison with one human's grasp of humanly mortal time has prompted many thinkers to contemplate the balance between a life lived and a lifetime of knowledge shared with others. Theorist and writer Murray Bookchin (1921–2006) devised the term 'social ecology' to embrace the shift as well as the duality in the individual who is both distinctly singular and also part of a larger shared collective (Bookchin 1982). His theory was based on the premise that 'humanity's continual *struggle* with its imagination lies at the very heart of the tensions within early society, which in turn has historically led to varying degrees of rational self-understanding as well as frightening, often atavistic regressions' (Bookchin 1996).

Nan Shepherd would have agreed with this, having clearly stated her view that her personal experience was written to share outwardly in a message to a wider, collective appeal. She felt most alive in the mountains, but sharing her experiences through writing meant that her knowledge could serve others.

Undoubtedly, there are limitations to some first-hand experiences, and among these are the inability to see the microscopic (or indeed, galactic macroscopic) without the aid of tools; the more-than-human equipment to sense geologic movement or Deep Time; and the inability to easily traverse various terrains, such as wetlands. This has the subsequent effect of devaluing these places, or foci of attention, and has subsequently led to disastrous consequences. Flawed human perception, precisely due to lack of first-hand experience, has lain behind many of the mistakes – and ignorance – of the true value of many environmental species and conditions. An example may be found in the dismissal of bogs. Often seen as 'wastelands', 'empty', and as land of no value, bogs have been drained, cut for peat, harvested for sphagnum moss and other plant life, and repurposed for Sitka pine plantations. This has come, it has been discovered, at a tremendous harm to this environment, as wildlife has declined, and carbon stored in the wetlands has been released back into the atmosphere through these actions (Hurd 2008). Now, for example, the regeneration of wetlands is seeing marked improvements, not only in the revitalised wildlife and insect world, but in the restoration of improved air qualities, enhanced carbon sinks, and greater flood management. But what these consequences demonstrate is the need for considered and careful restoration, conservation that is mindful of wider effects, and judgements that are made for planetary well-being, not just local economic impacts. Actions are needed, but the fullness and care in judging which actions to undertake are needed as well.

Response and Actions

Faced with the mounting evidence of human mistakes, and unsustainable consequences in the imbalanced environment, one cannot help but wish to respond with urgency. However, one also hesitates, pondering the limits of an understanding about being-in-the-world that is based on faulty conceptualisations, or arrogant assumptions about hierarchies of being. Murray Bookchin (1996) asked: 'What is nature? What is humanity's place in nature? And what is the relationship of society to the natural world?' This is another way of asking the same question about the dual role of being a single organism in a complexly interconnected world, and at the same time, one of billions in the species residing in a finite planetary environment. He wrote (Bookchin 1996, n.p.):

> In an era of ecological breakdown, answering these questions has become of momentous importance for our everyday lives and for the future that we and other life-forms face. They are not abstract philosophical questions [...] The definitions and ethical standards with which we respond to them may ultimately decide whether human society will creatively foster natural evolution, or whether we will render the planet uninhabitable for all complex life-forms, including ourselves.

These basic assumptions that both Bookchin and, more recently, Donna Haraway question are existential 'attitudes' a priori, which have profound consequences depending upon their construction. They are important for nothing short of life on the planet. And they also implicitly call for actions – first, in formulation of the requirements for Being (always a condition of *being-in-place*), and secondly, in the actions taken to adjust the insupportable. 'Action potentials' are what some researchers call the signals between organisms. Contemporary philosophers have also referred to these as 'agential intra-actions' (Barad) or 'vibrant matter' (Bennett). In the case of mycelial researchers, conduits of information between species, underground, are evidenced as agents for growth, inter-species communication, and the necessary mode of enmeshed organic life. In Hampton's video, she says (2022, 0.50–0.57): 'Fungi is a kingdom in itself. It re-embroiders death into life, weaving kingdoms together'.

Recognition of the entanglement of the human and more-than-human world also demands the understanding that communication is one more part of the inter-relationships. Communication has been established between individuals within non-human species, across species and biological kingdoms, and across geographical regions. Among many examples detailed in recent publications are the communications carried among trees with pheromones, sugary sap, and underground fungal ('mycorrhizal') networks (Wohlleben 2017), between various plants and the bees that pollinate them (Kimmerer 2013a), and in the Niassa National Reserve, Mozambique, between humans and honeyguide birds and the honeybees who sustain them both (Bridle 2022, 144). Symbiotic communication is not new, and many traditional peoples have observed patterns of more-than-human behaviour, and then adapted their own to become part of the ecosystem of mutually beneficial actions.

Even before humans make a decision to take action in their environment, perhaps the first 'action' made in response to living conditions has *already* been taken, as contemporary ecologists would suggest. These are the collaborative, cross-species responses, using many modes of communication as the examples above have indicated, and which are the channels for mutually ensured survival. Sheldrake's wonder at the symbiotic partnerships which fungi form with other species, and their own ability to form widely enmeshed networks, has been illustrated in Hampton's film and in the Figure 3.4 example. Communication and collaboration, then, are essential parts of the enmeshed world, the subject of ecosophy and science alike.

Poets and artists ask many of the same questions as philosophers, social theorists, and mycologists, but use different methods, and additional questions, to focus on these fundamental questions of being. What poets and artists are able to contribute to this consideration of the 'desire to know' is to link multiple modes of knowing in their art. *Being-in-the-world* affords a human the opportunity to be fully immersed in the totality of their environment, and it is this, 'the background from which all acts stand out' as Merleau-Ponty says, that prompts a complex response.[5] Beyond the physical reaction of sensation and perception, attendant emotions, memories, thoughts, and desires abound. It is precisely these qualitative responses that are the appeal to poets and artists. Their works move beyond the 'what' of experience, the cognitive awareness of actual phenomena, to incorporate the multivalent 'how', the qualitative, felt response.

In the following section, examples of artistic modes of enquiry demonstrate how the desire to know includes addressing the affective, qualitative aspects of lived experience, including characteristics of the intuited world, the numinous, and ineffable emotive responses. Poetry, film, art, music, theatre, and dance are forms which have the power to speak deeply to our shared conditions of human existence. These modes of communication appeal to some of the ways we recognise consequential effects of events and aspects of life we share with others in our species, moving beyond words in many cases to felt, internalised and embodied knowledge.

A Case for Hope

> Hope is the thing with feathers
> That perches in the soul,
> And sings the tune without the words,
> And never stops at all
> (Dickinson 1891, 27)

These lines by Emily Dickinson position the quality of hope as integral to human existence ('perches in the soul'). While one might recognise that there is an undeniable weightiness in the proof and legacy of human error, it is also inescapable; this current age of information provides ubiquitous evidence of human-made mistakes and unsustainable actions. Lived experience is the quality of knowledge that Shepherd values most in her *Living Mountain*, precisely because it is suffused with affective states in the wholeness of first-hand wisdom. Her writing does not seek to erase emotive qualities, but to embrace them as inherent to knowing. So, too, does Dickinson's 'hope' metaphorically sing in our soul with a 'tune without words'. Like birds ('things with feathers') who gave the poet a more-than-human physical shape for her imagination, the power of art and poetry finds its way directly to our comprehension of lived experience. For example, a counterbalance to the feeling of despair that might arise from the evidence of flawed human actions so readily apparent in the current climate crisis, may be found in the art and words of artist Olafur Eliasson (2019), who says: 'Every single little thing we do has consequences, and to recognize that doesn't mean we have to be dystopian or negative'. He continues:

> I understand the relevance of the fear-based arguments and narratives that are being thrown around, because they fill up our pool of worries to the extent of taking away our ability to dream of a positive future. But without positivity, it's very hard to create a consensus on the fact that we need to solve the climate challenge.

In his art, Eliasson seeks to find or create positivity, to renew a sense of wonder, and to focus on the value of conscious actions in opposition to despair and negativity. For him, being-in-the-world, and performing the desire to know, requires heightened attentiveness to natural phenomena in the environment. This is the first step in understanding what the changes are, and pointing to what needs focused action, in an imbricated ecosystem (see Figure 3.5).

Haraway, like Eliasson, embraces heightened awareness and 'staying with the trouble' (Haraway 2016). Her conceptualisation of the age is not that

Figure 3.5 Olafur Eliasson, 'Ice Watch' at the Tate Modern.
Source: Photograph by Alex Lentati/*Evening Standard* (2018).

of the Anthropocene, but rather what she has termed the 'Chthulucene', a period in which human and non-human must make together ('sym-poiesis': literally, making-with, together). This symbiosis is essential for Haraway, in contrast to the solely human-made and humancentric, which she rejects as an unsustainable environment. Both Eliasson and Haraway advocate handing over – or more properly – *recognising* co-dependence and interconnectedness as an enlightened concept of being-in-the-world, valuing differences as intrinsic to survival. Eliasson (2019) believes:

> "Often we make the mistake to think that having said something is closely related to having done something, and at some point we have spoken about it so many times that we actually feel we did it," he says. "One idea I think needs to be explored on a granular level is the simple step-by-step – how do you go from thinking and talking into doing and acting?"

Eliasson's question is to the point; how does one translate cognitive and sensory awareness into the motivation to act? One way is to build upon observation of the natural environment in creative, imaginative practices that produce works which will help viewers or readers to see freshly, to see anew. In *The Wild Iris* poem by Louise Glück (1992), the cyclical nature of the seasons, with winter's deep slowing, like a kind of death, or deep depression, but implicitly figured as an iris tuber ('consciousness/buried in the dark earth'), opens a space for regeneration ('returns from oblivion'). The poet's omniscient voice, speaking on behalf of the tuber, says '...that which you call death/I remember'. The voice remains after 'death', – a death which is not wholly a death – and carries a suffering in remembrance of life: to 'survive/as consciousness/buried in the dark earth'. This is a fear that many people have, to speculate some sort of lingering awareness in a state where there is only darkness, no movement, and seemingly no hope. And yet, there is a return to life, a cyclical rebirth, in which 'whatever/returns from oblivion returns/to find a voice'. The return of life is endowed with its own voice, to speak again, its own agency to *be* once again. The poem moves us with its metaphorical rebirth, the flowering in 'a great fountain', joining other wild irises. The final stanza juxtaposes the writer's own presence ('the center of my life') with a richly coloured imagined rebirth, the colour of iris blooms. The regeneration is a transformation from 'the other world' of 'dark earth', and being 'unable to speak', back into this one. Overlaid are the wild irises of the poem's title, the author herself, and us by suggestion, renewed after a period in the dark and life energy itself, cyclical, with slow periods of sequestration leading to a period of new growth and blossom.

In Figure 3.6, Potapoff's vision of a dark abstract organic form inhabits an ambiguous dark space that parallels Glück's 'consciousness buried in the dark earth'. The drawing is deep, powerful, and feels like a living energy existing in a vibrant space. The pale form in Potapoff's drawing is both in, and a part of, the darkness that holds it, just as metaphorically, Glück's buried consciousness is like an iris tuber, which needs the rest to store the energy it needs to blossom again. Cyclical movements of energy from high to low periods are requisite to each other.

Observation of the natural world that reconfigures in art, providing us with an inspirational metaphor, is similarly framed in the art of James Turrell (see Figure 3.7). In his various constructions called *Skyspaces*, he creates spaces that form interiors with apertures in the ceiling through which the sky is literally framed. The quiet, minimal space of the interior works as a three-dimensional 'frame', promoting a feeling similar to a chapel, wherein one wants to speak in whispers, or remain silent in reflective attentiveness. One's eyes are drawn upwards, to the moving clouds, to the flight of a bird, or simply to the expanse of sky itself. Spending time in a *Skyspace* is like a lesson in living; it focuses the mind on being, on living and life itself, as a primary state, undistracted by other things. One's breath slows to a deep peaceful rhythm. Shepherd would agree, having experienced something like this in her beloved Cairngorms. She states (2011, 6): 'Place and a mind may interpenetrate till the nature of both is altered'.

Figure 3.6 Katie Hart Potapoff, *With an Attentiveness* No. 2, graphite.
Source: Drawing by Katie Hart Potapoff (2022).

Figure 3.7 James Turrell, Two *Skyspace* constructions: Left: Roden Crater, *Crater's Eye*. Right: Roden Crater, *East Portal*.

Source: Art by James Turrell (2010). Photographs by Florian Holzherr.

There are moments when a mountain climber loses their orientation, and becomes uncertain, or lost. Or at other times, the climber sees differently (Shepherd 2011, 12):

> This changing of focus in the eye, moving the eye itself when looking at things that do not move, deepens one's sense of outer reality. Then static things may be caught in the very act of becoming. By so simple a matter, too, as altering the position of one's head, a different kind of world may be made to appear [...] How new it has become!

The contrasts between what one sees as habitual, predictable, and known environment, and the internalised landscape – the vision in the mind's eye – can be a startling and disconcerting effect. It is also the uncertain ground that may foster creative responses, seeking to represent the difference between these states. Dickinson's poem *The Lost Thought* (Dickinson 1896, 37) is a case in point:

> I felt a cleaving in my mind
> As if my brain had split;
> I tried to match it, seam by seam,

> But could not make them fit.
> The thought behind I strove to join
> Unto the thought before,
> But sequence ravelled out of reach
> Like balls upon a floor.

In this 'cleaving' the difference between illusion and reality, the jarring gap between the phenomenon and 'the thought behind', cause a disruption, likened here to balls of yarn rolling haphazardly upon a floor. In other words, the poet describes the separation between the objective and subjective experience, declaring that she 'could not make them fit' together.

These words, and the meaning behind them, are reiterated in the steady but understanding Shepherd when she observes (2011, 40):

> Thin mist, through which the sun is suffused, gives the mountain a tenuous and ghost-like beauty; but when mist thickens, one walks in a blind world. And that is bad: though there is a thrill in its eeriness, and a sound satisfaction in not getting lost. For not getting lost is a matter of the mind – of keeping one's head, of having map and compass to hand and knowing how to use them, of staying steady.

The knowledge of place, in other words, requires skills using both sensory and perceptive understanding, decoding what embodied experience has conveyed, and merging in a process of interpretation layers of human qualities which include emotion and memory, expectation and narrative in the sense-making process. Shepherd's way of 'staying steady' is to move beyond reliance on sight in the moment's 'thick mist', and to combine the many other ways of knowing her mountains. Muscle-memory, sounds, and her past climbs inform her knowledge. This multifaceted practice combining modes of attention, recall, imagination, and learning, is replicated in the ways that artists and poets combine associative and qualitative experience in their creative processes. What the work of artists and poets offers to the general public, and to scientists whose experimental methods exclude emotion, intuition, and associative connotations, is the ability to freely combine multiple modes of knowing and meaning-making.

Conclusion

In narratives of embodied experience, artistic and poetic creations, scientific studies, geopoetics, and ecosophical theories, we have considered qualities of being-in-the-world, of being driven by the 'desire to know'. We humans move with and among other life forms, carrying within our bodies hosts of other organisms at all times, treading upon ground through which messages are being sent as we walk. Birdsong as we move tells us more than incidental background noise. The more-than-human world is fully aware of our presence; they are communicating with us if only we are present to them and willing to listen. And returning home, we share stories of what we have seen, where we have been, and how it was.

And in turn, each constructed narrative affects the environment in myriad ways: these stories and images subsequently determine further human actions, in exploiting, conquering, conserving, or restoring sites such as mountains, rivers, forests, or moorlands. Perceptions shared, therefore, are forward-pointing, setting the stage for actions of those aiming to further know the nature of places. The ability to learn from the stories of one individual is the invitation to use imagination, to see in the mind's eye the qualities and circumstances, the markers, and messages of life shared with other forms of life. The path we know beforehand is conveyed through the feet of the storyteller, forester, mycologist, philosopher, and artist – each of whom showed us how things might be, as they have discovered in their journeys. And of course, in ways of Being, in the mountains, in the world, as Shepherd reminds us, 'the elementals play their own ungovernable roles'.

Notes

1 Nan Shepherd (1893–1981) first wrote *The Living Mountain* in the 1940s, as a personal reflection on the experiences she had while walking in the Cairngorm mountains of Scotland. She did not publish this book until 1977. The reference citations used here are from the 2011 reprint by Canongate.

2 These writers include Mark Boyle, Roger Deakin, Annie Dillard, David Gange, Kathleen Jamie, Richard Mabey, Robert Macfarlane, Alastair Moffat, Alice Oswald, Amanda Owen, Merlin Sheldrake, Amanda Thomson, and Peter Wohlleben, among a myriad of authors.

3 *All My Relations*, First Nation Pedagogy. https://firstnationspedagogy.ca/interconnect.html See, 'A Declaration of Spirit,' video, on this website.

4 'Deep systems' are a part of ecological systems theory (EST), and also known as human ecology, an ecological/system framework developed in 1979 by Urie Bronfenbrenner; James Lovelock and Lynn Margulis in the 1970s named 'Gaia' as the complex condition of inter-relations that the sciences had dealt with separately but that were more effectively conceived as one holistic state; 'earth system science' is articulated by Tim Lenton and his group at the University of Exeter, focusing on understanding the Earth as a system, modelling evolution, ecology, and biogeochemistry, providing early warning of climate tipping points, and identifying positive tipping points towards sustainability.

5 *Being-in-the-world* is a Heideggerian phrase, and one which emphasises the qualities of lived experience, and in living in the world, wherein one is constantly in a state of uncovering, disclosing other things and oneself. This phrase emphasises *being* as a process of meaning and growth, in which humans are co-participants.

Bibliography

Abram, David. 1997. *The Spell of the Sensuous: Perception and Language in a More-than-Human World*. London: Vintage.

Andreyev, Julie. 2021. *Lessons from a Multispecies Studio: Uncovering Ecological Understanding and Biophilia through Creative Reciprocity*. Bristol: Intellect.

Barad, Karen. 2007. *Meeting the Universe Halfway: Quantum Physics and the Entanglement of Matter and Meaning*. Durham, NC: Duke University Press.

Baugh, Bruce. 2022. *Philosophers' Walks*. Abingdon: Routledge.

Bennett, Jane. 2010. *Vibrant Matter: A Political Ecology of Things*. Durham, NC: Duke University Press.

Bookchin, Murray. 1982. *The Ecology of Freedom: The Emergence and Dissolution of Hierarchy*. Palo Alto, CA: Cheshire Books.

Bookchin, Murray. 1996. The Philosophy of Social Ecology: Essays on Dialectical Naturalism. Second Edition. Accessed 10 August 2022. https://theanarchistlibrary.org/library/murray-bookchin-the-ecology-of-freedom

Borthwick, David, Pippa Marland and Anna Stenning, ed. 2020. *Walking, Landscape and Environment*. London: Routledge.

Bridle, James. 2022. *Ways of Being: Beyond Human Intelligence*. London: Allen Lane.

Cox, James L. 2018. *Restoring the Chain of Memory: T.G.H. Strehlow and the Repatriation of Australian Indigenous Knowledge*. Bristol, CT: Equinox.

Dickinson, Emily. 1891. *Poems: Second Series*, edited by T. W. Higginson and Mabel Loomis Todd. Boston, MA: Roberts Brothers.

Dickinson, Emily. 1896. *Poems: Third Series*, edited by Mable Loomis Todd. Boston, MA: Roberts Brothers.

Eliasson, Olafur. 2019. "Eliason on Art, The Climate Crisis, and Why We Should All Be Activists." In *Highsnobiety*. Accessed 1 September 2022. https://www.highsnobiety.com/p/olafur-eliasson-artwork/

Glück, Louise. 1992. *The Wild Iris*. New York: Harper Collins.

Guattari, Félix. 2000. *Three Ecologies*. Translated by Ian Pindar and Paul Sutton. London: Athlone.

Hampton, Léonie. 2021–2022. "Our Body is a Planet." University of Exeter, MRC Centre for Medical Mycology, in partnership with Arts and Culture. Accessed 9 August 2022. https://thefungalthreat.com/take-part-and-discover/art-meets-medical-mycology/our-body-is-a-planet/

Haraway, Donna. 2016. *Staying with the Trouble: Making Kin in the Chthulucene*. Durham, NC: Duke University Press.

Hurd, Barbara. 2008. *Stirring the Mud: On Swamps, Bogs, and Human Imagination*. Athens: University of Georgia Press.

Kimmerer, Robin Wall. 2013a. *Braiding Sweetgrass: Indigenous Wisdom, Scientific Knowledge, and the Teaching of Plants*. Minneapolis, MN: Milkweed.

Kimmerer, Robin Wall. 2013b. "Council of the Pecans: Lessons on Native Resilience from a Venerable Tree". Orion: Nature/Culture/Place, September–October. Accessed 10 February 2026. https://orionmagazine.org/article/pecan-trees-indigenous-land-rights/

Lenton, Tim. 2016. *Earth System Science: A Very Short Introduction*. London: Open University Press.

Macfarlane, Robert. 2013a. *The Old Ways: A Journey on Foot*. 1st edition. London: Penguin.

Macfarlane, Robert. 2013b. The Living Mountain, podcast. Accessed 9 August 2022. https://www.bbc.co.uk/sounds/play/b03mfndd

Macfarlane, Robert. 2017. *The Wild Places*. London: Granta.

Merleau-Ponty, Maurice. (1945) 2002. *Phenomenology of Perception*. Translated by Colin Smith. London: Routledge & Kegan Paul.

Modeen, Mary and Iain Biggs. 2020. *Creative Engagements with Ecologies of Place: Geopoetics, Deep Mapping and Slow Residencies*. London: Routledge.

Nelsen, Melissa K. and Daniel Shilling, ed. 2018. *Traditional Ecological Knowledge: Learning from Indigenous Practices for Environmental Sustainability*. Cambridge: Cambridge University Press.

Nietzsche, Friedrich. (1908) 1969. *Ecce Homo: How One Becomes What One Is*. Translated by Walter Kaufman. New York: Vintage.

Sheldrake, Merlin. 2020. *Entangled Life: How Fungi Make Our Worlds, Change Our Minds, and Shape Our Futures*. London: Vintage.

Shepherd, Nan. (1977) 2011. *The Living Mountain*. Edinburgh: Canongate.

Tsing, Anna Lowenhaupt. 2015. *The Mushroom at the End of the World*. Princeton, NJ: Princeton University Press.

Tucker, Judith. 2020. "Walking Backwards: Art between Places in the Twenty-First Century." In *Walking, Landscape and Environment*, edited by David Borthwick, Pippa Marland and Anna Stenning, 127–146. London: Routledge.

Varela, Francisco J., Evan Thompson and Eleanor Rosch. 1992. *The Embodied Mind: Cognitive Science and Human Experience*. Cambridge, MA: MIT Press.

Walton, Samantha. 2021. *The Living World: Nan Shepherd and Environmental Thought*. London: Bloomsbury.

Whyte, Kyle Powys. 2018. "What Do Indigenous Knowledges Do for Indigenous Peoples?" In *Traditional Ecological Knowledge: Learning from Indigenous Practices for Environmental Sustainability*, edited by Melissa K. Nelsen and Daniel Shilling, 57–81. Cambridge: Cambridge University Press.

Willis, A.J. 1997. "The Ecosystem: An Evolving Concept Viewed Historically." *Functional Ecology* 11 (2): 268–271.

Wohlleben, Peter. 2017. *The Hidden Life of Trees: What They Feel, How They Communicate: Discoveries from a Secret World*. Glasgow: William Collins.

Wohlleben, Peter. 2019. *The Secret Network of Nature: The Delicate Balance of All Living Things*. London: Vintage.

4 Mire Art, Environmental Concern, and the Changing Cultural Heritage

Pauliina Latvala-Harvilahti and Kirsi Laurén

This chapter examines the changing use and attitudes towards mires, particularly from the perspective of mire arts and artists. The topic is studied by analysing interviews with Finnish artists who create mire-related art or art on mires. The study is based on the notion that there has been an increase in the number of art events and performances taking place on mires across Europe in the 2000s (Laurén 2008; Pirtola 2008; Green 2018; *Peatart Gallery – Treasures from the Bog* 2022). We understand mire arts as a continuation of the phenomena of environmental art (Thornes 2008; Gellatly et al. 2016). The popularity of mire art is explained by its timeliness in relation to people's feelings about their environment. Some artists create mire art just because they are fascinated and inspired by mires as a natural area, and some artists are motivated by both their personal and global environmental concerns. It is a common phenomenon that they take a stand for the protection of pristine mires by getting people to pay attention to threatened nature. This environmental and social activity is called 'artivism', which is a combination of the words 'art' and 'activism'. An artivist is an artist who makes work that reflects his or her activism. Artivism has a broad scope, which is also reflected in our research data. In this chapter, we focus on mire artivists and highlight how mire artivism can be attracted implicitly through soft artivism, and explicitly through strong artivism.

Interest in Mire Arts and Artivism

The main research question in this study is: In what ways has the environmental concern of mire artists affected their artistic work? With this question, the aim is to shed light on the broader issue of whether mire art is changing the cultural heritage associated with mires. Cherished and renewed in everyday life, heritage includes the human relationship with nature. Alongside cultural policy, all people have the possibility to define and participate in cultural heritage. As people have moved, worked, and used mires in many ways, mires are associated with tangible and intangible cultural heritage. To avoid sinking and wet places in the mire requires knowledge, skills, and experience of nature, and these are often passed on tacitly from one generation

DOI: 10.4324/9781003727972-4

to the next (Laurén 2006). In addition, this kind of intangible cultural heritage supports community-based practices and communication (on intangible heritage, see Kearney 2009; UNESCO). Local (mire) stories, names of the places, and other rich oral traditions and beliefs encourage a sense of identity and responsibility. Today, however, the cultural heritage of mires is more strongly experiential, eco-conscious, and future oriented (Latvala-Harvilahti 2022a; 2022b; Laurén 2022). Knowledge about mires, together with experience-based reasons to go to the mire, makes the human-mire relationship more sustainable. The mire art environment is not only a natural environment, but also a social and societal environment, the sustainability of which must also be considered from an artistic point of view (Kaukio 2022a, 2022b, 464). Thus, the boundaries between natural heritage and cultural heritage, and art and cultural heritage, have become blurred when talking about the human-mire relationship as cultural heritage.

Together with forests, mires and peatlands are a common habitat type in Finland, covering about one-third of the land area. In different cultures, these low-lying, wet, and sinking mires are often perceived as unesthetic and marginal landscapes that are difficult to walk on and navigate (Lehtinen 2000). In addition, anomalous mires formed by a mixture of soil and water have given rise to mythical and frightening images, where it is possible to sink into a mire, get lost in it, and even encounter supernatural creatures. Mires symbolise the unconscious and are associated with stages of death and horror (Wilson 2005; Laurén 2006, 98–108; 2011, 108; Laaksonen 2008; Giblett 2014, 11; Joyce and Staunton 2020, 85). As a result, there has long been a negative perception of pristine mires, and efforts have been made to transform them into something more beneficial to humans. This has particularly been the case in countries with a lot of mires, and mires have been drained for corn fields, forests, and peat production. Thus, in Finland mires have represented both hard work and everyday survival (Laurén 2006).

Mires have always been a part of the shared experience of Finnish 'Everyone's Rights', which encompass the right to walk, ski, pick berries, and camp temporarily in nature. For outdoor activities, the mires are considered to belong to all people, even though they might in fact be owned by someone. The owner of the mire or forest can be the state, the private sector, a family, an association, or even a church. In southern Finland, mires are mainly privately owned, while in Northern Finland they are mainly owned by the state. The number and conservation status of pristine mires are unevenly distributed, being worst in Southern Finland (Hyvärinen et al. 2019, 50–57; Everyman´s Rights 2000). In protected areas, the Nature Conservation Act allows for restrictions on this to achieve the conservation objective of the area, and activities may be restricted if this is necessary to preserve the fauna or flora of the area[1] (Nature Conservation Act 1996).

Today, it is known that mires and peatlands are globally important stores of soil carbon, a sink for carbon dioxide, and stores of atmospheric methane (Strack 2008, 9). The effects of mire drainage and climate change also threaten the source of information about past cultures and environments. Consequently, in addition to their ecological functions, mires have intangible human aspects like personal attachments, and aesthetic and recreational meanings, which are important parts of the value of mires (Gearey and Everett 2021). Preserving pristine mires is a way to prevent habitat loss, mitigate climate change, and protect the emotional values and cultural heritage that are important to humans. Therefore, there is a growing interest in protecting mires and peatlands, and preserving their biodiversity. Nowadays, mires are also increasingly seen as areas of aesthetic and recreational importance, and where nature used to be a place for solitary wanderers, the trend nowadays is to be social in nature (Sorakunnas 2022). Additionally, artists have also noticed the value of mires, and thus the role of artists and art in the current global climate catastrophe has become intensified.

This chapter seeks to examine the views and experiences of artists in the light of the concept of *environmental concern*. In addition, the analysis is completed by using the concept of *place attachment* which helps to reflect the connection between the human–mire relationship and individual concern that is felt towards mires. This chapter is structured as follows: First we present the research material and analysis methods used in the study. The research is then followed through key theoretical and analytical concepts of environmental concern and place attachment. We then move on to the analysis of the research interviews, and finally present the conclusions of the study.

Research Materials and Methods

This research, and the research material collected in it, are part of a larger Mire Trend research project (2020–2023, University of Eastern Finland) that studies the change in the cultural heritage associated with mires in the twenty-first century.[2] The empirical research material of our study in this chapter is based on 33 thematic interviews conducted between 2020 and 2022. The artists selected for the study were mostly found through their exhibitions, books, performances, personal websites, and social media. Around two-thirds of respondents are women, and the rest are men. Their age range is between 25 and 65 years old. Most of the interviews were individual, but there were a few group interviews with two or three interviewees. Most of the interviewees represent the sector of visual arts (photographers, painters, video artists), but also performing arts (musicians, dancers, theatre), arts and crafts (a textile artist), and self-made life artists (ITE).[3] Some of the interviewed artists represent both the academy and the arts. Most of the interviews were conducted individually and face to face, and some by phone due to the Covid-19 pandemic. The average length of an interview was an hour.

The interviewees participated in the research on a voluntary basis, and they signed the archiving permission for the Finnish Literature Society's Archive. This means that the material will later be available to other users, in accordance with the rules of the archive. The interviewees were given the opportunity to set restrictions on the use of their interviews. Few imposed restrictions, and none requested anonymity. However, in this chapter the interviewees are not identified by name, and instead, the interviews are referred to by consecutive numbering (I1, I2, I3, and so on, where I = interview, and 1–33 = interview identified by number). The artists we interviewed represent different fields of art and some are more widely known than others, which is why we do not use their names in this study. It is not the works of art as such that are the subject of our analysis, but rather the artists' way of verbalising their thoughts in interviews. Although the anonymity of the artists is not perfect in the study (the names of the works of art reveal their creators), for the matter of equality, we do not refer to them by their own names. The interview questions were organised in three themed sections: (1) Mires in mental and physical landscapes and current opinions on their use; (2) Mire and art; and (3) Environmental crisis and art. Themes such as the human-mire relationship, drainage of mires, various forms of mire use, and the importance of the nature of the mires were central to the interviews.

As a method for analysing the interviews, we use qualitative theory-based content analysis. This means that first, we studied the interview material we collected ourselves and made observations about the key themes and issues that are important to the interviewed artists, and which answer our research question. The main emergent themes are concerns about the state of the mire environment and the artists' own experiences of the mire. On this basis, we chose the theoretical concepts of *environmental concern* and *place attachment* to explain and open the research material, and to guide the analysis. In our analysis, we consider both the explicit and implicit meanings expressed by the interviewees about how the experience of environmental concern affects them and the art they make.

Theoretical Dimensions of Environmental Concern and Place Attachment

Globally, humans have significantly altered natural environments such as water bodies, forests, mountains, mires, and peatlands, leaving visible and often irreversible traces such as pollution and loss of habitat. Compared to other species on Earth, humans have the greatest impact on natural resources. People respond to environmental problems differently, influenced by both personal and social factors. Personal factors include, for example, childhood experiences, values, and place attachment, whereas social factors include issues such as proximity to problematic environmental sites, norms, and urban–rural differences (Gifford and Nilsson 2014).[4]

Of the above categories, personal factors are the most important for this study. The quality and intensity of environmental concern depend on the attachment one has to different places, and in this study to mires. Thus, when the focus is on environmental concern and the activities that arise from it, the interest is mainly located in the *place attachment* of the people. Mires form a valuable site for individuals who wish to take mires into account by the means of art-related projects or performances. But regardless of their location, artists are surely not 'outsiders' in feeling current climate emotions such as grief, sorrow, and anxiety.

The articulation of environmental concerns related to the mires requires an acknowledgement of the cultural meanings of the interviewees' individual place attachment. Simply put, this refers to the emotional bonds that form between people and their physical surroundings. Place attachment includes the feeling of continuity, in other words that the time levels of the past, the present, and the future are considered important. There are various notions of place attachment, depending on the subject and the method of research. For example, place identity, place dependency, and the relationship to nature are combined in multi-dimensional place attachment (Williams and Vaske 2003; Manzo and Devine-Wright 2021; Hernández et al. 2014; Raymond et al. 2010; Segers et al. 2021). Environmental philosopher Shierry Weber Nicholsen (2002, 46) has written: 'Concern is the effort to make oneself worthy of what one loves. It is expressed in working to know and appreciate the beloved and to preserve its value'. Environmental concern arises from the feeling and perception that some people see the damage done to natural environments, while others do not. Accordingly, the efforts to raise awareness of the dangers threatening, for example, fragile mires reflect a hope for the future: we are not alone, and the saving and healing of nature is possible (Nicholsen 2002, 163).

According to a recent survey, 79% of Finns are concerned about the state of nature in the world. The survey also showed that Finns feel that being out in nature helps them to manage the difficult emotions evoked by climate change (*Suomalaisten luontosuhteet* 2021). As a country with a high level of education and GDP, Finland's citizens are aware of environmental threats, and prepared to take action to prevent them. However, not everyone acts in favour of the environment (Johansson Sevä and Kulin 2019). Research on people's daily activities, ways of thinking and acting, as well as the changes that occur in them, has become increasingly important for addressing sustainable future issues and development, and community action and civic engagement in its various forms generate collective momentum and articulate citizens' environmental concerns (see, e.g., Häyrynen and Pynnönen 2020; Lähde 2022; Laakso and Aro 2022). It is also essential to be aware of the role of emotions in the human relationship with the environment.

Environmental Concern in the Implicit Soft Expression of Artivism

The intensity of expression of opinions and goal-oriented activism varies among artists. There is a gentle approach of soft mire artivism which tends to display the beauty and diversity of mire nature, without any obvious features that can be considered to encourage activity. Regardless of where on the scale artivism is placed, those who experience art can be emotionally affected

(Golanska and Kronenberg 2020; Latvala-Harvilahti 2021; see also Rodal et al. 2019). For many artists, the most important value of artwork is not to convey political messages, despite the fact that works were often inspired by individual or community-based environmental concerns perceived by artists. This point of departure reminds us that even though art may widen understanding or change people's opinions on nature values, politicians are responsible for legislation. In the research interviews, the artists talked a lot about mires and their personal relationship with them and with nature in general, and not just about their art and its goals. The representation of the beauty of nature is seen as an important emotive factor, even if the starting point for the art is not to make a statement. This is how a photographer and visual designer put it:

> I wish I could convey emotions through my stories and make people think about something, especially emotions. The focus in my nature-related works has been on beauty. I don't really take a stand. [...] But we are so distanced from nature that we are destroying it without knowing it, everyone should wake up and realise this. This earth is not just ours.
>
> (I28)

An artist may have the desire to make a statement, but personal artistic visions can prevent them from doing so. Alone, an artist may not want to take a public stand, although when working with other artists, it can be meaningful because it makes it possible to bring together different perspectives (I21). An example of a participatory mire art event organised by a group of artists together with other actors has been the *Mire Reception* [*Suovastaanotto*], held in a remote mire in the Kainuu region in the summer of 2020 as part of The Mustarinda Association's residence activity (see Figure 4.1). The goal of the association is to connect art and science, and to promote the ecological rebuilding of society. Artists representing visual art and sculpture explained that the aim of the event was to get villagers, municipal politicians, landowners, and other invited guests to share the experience of mires, to celebrate the intrinsic value of them, and to discuss different perspectives and preferences related to mires. The event followed a festive format, with food, speeches, and art performances. The event was not intended to be high-brow and enlightening, but rather warm, uplifting, and community based (I10; I6; I33).

By organising participatory environmental art events, artists work with people to address affective concerns. Reflecting on local involvement in safeguarding mires, it is important to remember that historically, mires have been used in everyday purposes. For example, in the Viiankiaapa mire in Finnish Lapland, fishing, finding grass for animals, hunting, and berry picking have been part of local life for generations. Local history and the values of the

Figure 4.1 Participatory *Mire Reception* providing art and mire-related information attracted local people to the mire in Puolanka municipality, Finland.

Source: Photograph by Virpi Kaukio (2020), used by permission.

mire itself are transmitted through oral history. These aspects of heritage form part of the local identity and well-being, although their importance is not always recognised (cf. Flint and Jennings 2021). The above-mentioned viewpoints reflect the values of the Viiankiaapa mire at the local level. Mires have formed, and for many, still form a key place attachment. For these local people, the mire is not just a defined area on the map or a protected area, but rather it is part of their home region and heritage. Visible coverage of these events has brought people's concerns and experiences to a wider audience in a way that does not get bogged down in the confrontation or ready-made discourse typical of environmental projects (I29).

Art can influence the values and attitudes associated with nature in many ways. One of the contextual factors is the kind of mire environment the artist chooses for the audience to experience. Endangered mires were sometimes chosen as a site for art performances because of the artist's own experiences of eco-anxiety. This is precisely why localism as a personal relationship with nature is likely to raise concerns about the state of nature in general. Another reason for the importance of localism for environmental awareness is that being close to nature in a figurative and tangible way simply makes people aware of changes in the environment. This may explain why the artists we interviewed often see the mires as independent actors, and while mires are part of the forest environment, they differ from it in the way they sound, smell, look, and feel. Furthermore, the agency of mire also refers to the non-human actors in the mire, and thus emphasises the co-existence of different species. Thus, in the light of our research material, the independent agency of nature and its various elements is particularly prominent in environmental artivism. Many mire artists' understanding includes a co-agency of mire, which is close to the bio-artists' idea of using living organisms and other living material in art (Pietarinen and Qureshi 2023; Härkönen et al. 2023). Mire art also includes works in which the changing mire's material environment is part of the art. In this case, the co-agency of the mire also challenges the question of who controls the work, human or non-human actor (Härkönen et al. 2023).

In the interviews, mires are described as a wilder and more pristine form of nature than forests (I27; I7; I3). The independent nature and agency of the mire is emphasised in the photographer's interview, in which she describes a spring visit to the mire and heard its heart beating:

> It was a warm day, and the mire was beginning to melt. Mud had formed, melted and frozen pools on the surface of the moss. There was an incredible bubbling and gurgling sound. The sound changed, making me feel as if the mire was alive. I could imagine that the heart of the mire was beating. It was a truly wonderful experience.
>
> (I27)

The agency of mires is highlighted in photographic art and in the interviews with photographers (see Figure 4.2). The photographers are often very close to the mire they are photographing. Particularly when taking macro photographs, they sit or lay down in the mire, which also gets their clothes wet. At ground level, the mire is felt on the skin and smells strong (I22; I3; I32; I30; I18). The mire's own agency is reflected in its unpredictability; it is different depending on the time of year, the time of day, and the conditions of the weather. When visiting a mire, the photographer cannot know in advance what kind of photographs will be possible. Intuition guides their photography, and the photographer must adapt to working with the mire, 'breathing in sync with it' (I5). To get good photos, it is necessary to know how to settle into the rhythm of the mire and to live with it. The unpredictable and agentive nature of the mire is also revealed in the interviews with dance and performance artists performing on the mire. The specific nature of the mire means that the artist must perform and move within the boundaries of the mire, and together with it (I12; I19). Observing and understanding the autonomous agency of the mire brings the artists' attachment to place more to the fore.

Video installations are made as a close-up journey which presents nature and multiple time scales from the perspective of the mire, and as a co-existence

Figure 4.2 Photographing the details and landscapes that capture the character of the mire requires time and knowledge of nature. Photo exhibition *Torronsuo – magical light and enchanting scent*, Villa Elfvik, Espoo, Finland. Photograph by Kirsi Laurén (2021).

of humans and non-human animals (I14). This leads to an arousal of environmental concerns, and a desire to protect mires by taking an implicit stand with their art. Artists' comprehensive relationship with nature includes more-than-human aspects that support resilience among the current wretchedness of environmental problems. Sustainability issues and how to address them have become more prominent in the arts and cultural heritage (Siivonen et al. 2024). The aspects of heritage of the mires include expected silence (from human infrastructure such as traffic), and instead, the soundscape is understood to be made up of the sounds of nature; both animals and, for example, the sounds of weather such as wind and rain. For individuals and communities, the living heritage of the mires is rich not only in the form of sensory perceptions, but also for other senses and emotions that are affected by mire art.

However, a photographic exhibition on mires does not always take a direct stance in favour of them. Rather, the artist may be motivated by a desire to highlight the special beauty and intrinsic value of the mire through her or his exhibition. In this way, by sharing their own aesthetic experience publicly for others to appreciate, the artist implicitly, and without underlining the fact, sides with the mires. Importantly, in the exhibitions, the mire photographer brings out things that others cannot see. In this way, the photos bring out the independent agency of the mire, which is manifested, for example, in different atmospheres and details that are usually beyond human perception.

Such photos are captured at unusual times, such as very early in the morning or at night, when people are less likely to be out in nature. So, the exhibition helps to open the eyes of others (I22).

The use of the digital environment and the benefits of using technology to access the feeling of nature and cultural heritage are reflected in diverse ways in the case of mire art and its implicit soft expression. As examples, the online mire videos range from serious to humorous. A virtual artwork using augmented reality technology provides information on mire restoration to the visitor. In addition, people with reduced mobility can experience the mire virtually at a distance (I16). YouTube, on the other hand, provides a global audience and options for local performances. For online audiences, individuals have carried out distinctive, even peculiar artistic actions in the mire. By using a hunting camera 'The Freak of the Cloudberry Mire' adventures alone and almost naked in a north Finnish mire. The two videos are freely available on YouTube. The only 'clothes' the Freak wears is a 'kind of facemask' as underpants and part of a sewer pipe on his face as a didgeridoo. The Freak walks on the mire, appears and disappears, and has a message for people in the time of climate crisis. Ringing a cowbell, he returns, picking up a plastic bag left in the mire. On noticing it, he starts to mutter and looks angry. The self-taught artist who made the video is in favour of protecting the mires, and says the character he created has a stance (I9).

As described in the work above, there are cultural traditions associated with mire art which can be country specific. In Finland, the history of mire art includes naked performances and photos of naked people. Especially in the 1970s–1990s, art exploring the relationship between humans and nature often showed muddy bodies in the mires. Photos of naked artists in the mire were taken even in wintertime. Nudity in mire art also suggests that by taking off one's clothes, one also removes hierarchies, so that everyone can take part in the performance (see Takalo-Eskola 2017). The legacy of nakedness in mire art has been continued in amateur environmental video performances. One of the professional artists interviewed was surprised to notice that the theme of nakedness is still present in digital material. For her, taking off their clothes represents abandoning sophistication (I20). Visual performances recorded in mire nature such as the case of the 'Freak of the Cloudberry Mire' form both pedagogical and humorous content for environmental concerns. In addition to the above-mentioned platforms, social media is used to invite people to participate in art projects, where for example, a dancer invites others to dance on behalf of the mire. In such cases, participation then becomes a public statement, and artists and the public can both join in the social debate on mires (I24).

For some mire artists, the mire is a natural habitat where they have lived all their lives. This is the case for a creative self-made artist (or 'tinkerer') who built a cabin and sauna on the pristine Ruuhineva mire owned by himself (see Figure 4.3). The buildings and the mire for peat baths in front of the

Figure 4.3 The Kammi village [Kammikylä] and its sauna, built by a self-made artist on the Ruuhineva mire, Finland. The sauna is open to all visitors.

Source: Photograph by Virpi Kaukio (2020), used by permission.

sauna are freely accessible to anyone, so passers-by can take a sauna during summertime. In this way, he wants to offer people who do not know each other the opportunity to meet and, importantly, to discover the beauty and value of the mire, and enjoy the peace and silence of nature. In his view, this is what today's busy people need (I4). At the same time, both the sauna and being in the muddy mire provide the naked body with the physical contact of the mire. This bodily contact reinforces the place attachment associated with the mire.

Environmental Concern in the Explicit Strong Expression of Artivism

On the impact of environmental concerns on mires, the most pronounced objective is to engage people in transforming society in response to the climate crisis. We call this strong mire artivism. Mire artists' collaborations with nature specialists (e.g., scientists), eco-activists, volunteers, and residents can be strong. Artists and other agents together provide a sense of hope and foster individual engagement and 'future agency'. This concept refers to artists' consciousness of pro-environmental art, and a willingness to act for a sustainable future – awakening others in everyday life to take a stand for mires and nature (Latvala-Harvilahti 2021). Recent research has also sought to promote a more comprehensive understanding of the complex relationships between humans and animals in the context of artistic expression and the pursuit of artistic sustainability, particularly where the value of all living beings and their role in the realisation of sustainable artistic endeavours is recognised (Hernandez Cervantes and Huhmarniemi 2023).

Many artists who depict mires in their art or perform there have found local mires to be personally important. A personal attachment to place and earlier experiences of nature motivate these artists' art projects. Such an affective human-environmental relationship includes a place attachment that highlights the experiences of beauty, feelings of well-being, as well as a love of home or 'longing for roots' that often goes back to childhood (Milton 2002, 64; Torvinen 2012). Artists who have a very close and caring relationship with mires can be called 'mire lovers', and their relationship with mires has typically developed by living close to them and getting to know their way of life. For those artists, the mire is like a home that offers intimacy and a place where they can be themselves and free their imagination. For them, the natural world of mires is entwined with the sense of home and the sense of self (Nicholsen 2002, 40–41).

For a Finnish award-winning writer specialising in nature photography who has lived among mires and forests since childhood, the key experience of becoming an active nature conservation activist was the large-scale, systematic drainage of mires in his neighbourhood that took place in the 1960s. He has published several books of photographs and writings (e.g., *Kurkimaa* [Crane land] (1980) and *Tansseja aavalla* [Dancing on the Aapa mire]

(1994)) defending pristine mires and has held several photographic exhibitions. He has always opposed the intensive exploitation and destruction of nature for economic interests, and mires and environmental concern made him a photographer and writer:

> After getting a camera and a typewriter, I found a law-abiding way to work for nature. As soon as the drainage of the mires began, I opposed their drainage with all my heart and soul. From the age of 18, all through the late 1960s, I was writing in every possible newspaper opposing and criticising the drainage of mires illustrated with photographs I had taken. It was then that I became a photographer and perhaps also a nature writer.
>
> (I17)

However, a knowledge of mires does not always or necessarily mean that artists are specifically familiar with the ecology of mires, but more that their knowledge is based on their first-hand experience of mires. So, they have also become aware of the independent agency of the mires. Their love for the mire has grown and deepened the more they have spent time there, and incorporated the mire into their art. For example, for one artist who paints mires, the mires represent wildness, freedom, and solitude. She chose mires for her art because they reflect her own spiritual essence. Her concern for the mires grew as her personal mire attachment became closer. In addition to her paintings, she has written articles for newspapers, taking a stand for the protection of mires (I7).

Engaging with natural areas and making mire art involves controversial ideas and different perspectives, encouraging a strong stance. For example, the rise of intrinsically valuable nature photography with mobile phones may raise tricky questions about the origin of the minerals used by mobile phones, the number of pictures taken, and the need to use electricity. Thus, the contradictions that artists present in their art performances are not simple. Art offers a little seed of thought, but everyone interprets what they experience from their own point of view (I15). In addition to land use, melting glaciers and burning forests have given artists the impetus to immerse themselves in a new pro-environmental concern perspective. As an intangible cultural heritage and part of our relationship with nature, interaction with the mire is revealed in interviews and in the artworks as a reflection of our values. Mire art exposes people to a view and experience of nature that appeals to the emotions and senses, often emphasising that the nature of the mire is not a separate entity from the human being. But although it is easier to exploit nature that is thought to be separate, one artist reflected that Finns still largely think that nature can be used indefinitely:

> It's a bit like forests for Finns, we always imagine that there's plenty of them and we can pull what we want out of them, until one day

> it's the yard from which all nature has been taken away, and at that point, perhaps, people will wake up to the fact that it's not in a very good state.
>
> (I1)

But there is hope. She suggests that research into the role of mires as carbon sinks may have reawakened people. The interconnectedness between human activities and the life of mires is shaking up attitudes and traditions. This is especially true today when we are surrounded by news about species loss and the environmental crisis in a very different way than we were decades ago.

Many mires have intergroup tensions that are seen between state interests, mining company activities, and the contradictory values of residents and communities. In the interviews with artists, concerns for the future of mires emerge. At the local level, this is not a short-term concern, but concerns about the mire have persisted for up to ten years in areas with mining plans. Artists have recognised people's concerns about the loss of place identity. Under the pressure of mining, they can lose the landscape that is important to them (I11). This also explains why mires have their defenders – individuals who work on civic engagements and are partly organised in social media (e.g., 'the Guardians of Torronsuo mire' in Tammela, Häme region, or the 'Viiankiaapa Movement' in Sodankylä, Finnish Lapland). In our research materials, artists collaborate with movements and are sometimes members of them. An example of the strong place attachment and emotional bond to ex-homestead mire is reflected in an art event where an opera singer returned from abroad to perform in the Viiankiaapa mire. Her live musical ritual was held in the bird tower. In the interview, the strong expression was present even in the use of her language: 'We, the people of Lapland should defend our own areas and think two generations ahead – animals and plants instead of the present-day survival of humans' (I19). As presented above, the mires can evoke strong subjective reactions as meaningful places (see, e.g., Campbell 2021), for example, hopes that as fragile natural places, at least the most significant mires would be protected from mining. This kind of place attachment, which is seen in values and activities, reflects the relationship to mires as heritage. However, intangible cultural heritage can only be appreciated as heritage when it is recognised as such by communities or individuals (Kanerva and Mitchell 2015, 24).

By planning exhibitions on mires and climate change, artists have aimed to convert their own thinking, instead of moralising others. Artists recognise the importance of responsibility and shared future agency: 'These are our choices, whether we care about nature or not, it's not really that big a deal in the end' (I31). Environmental concerns are thus strongly linked to the future use of mires and peatlands. Concern about mires is reflected, for example, in the emphasis on responsibility, the fears of changes in nature, and the sad mentions of existing actions on mires as well as the protests they have engaged in through art in mires (I26; I1; I25; I2; I7). Artistic performances in non-natural mires also fall within the scope of strong artivism. Drained mires, mires in the process of restoration, and peatlands taken into energy production are examples of such areas. But although the use of energy peat has decreased, interviewees were still concerned about whether there is a timely and sufficient understanding of the need to protect peatlands (I2; I7).

The interviewees stressed that the mire represents nature, but that it has been and continues to be influenced by human actions. Artists' personal emotions and their values work as a motivator to create artworks that raise questions, but which also provide a sense of hope. During artworks and performances, specific places in the mires are transformed into shared spaces for new kinds of environmental dialogues between nature, the artist, and the audience (see Figure 4.4).

Whether presented outdoors or indoors, mire art has deliberately highlighted the value of mires in the questions of biodiversity and climate crisis. This can be seen in many stages of such artworks and projects. Firstly, it is present in the names of the art projects (e.g., the title *Dirge on the Mother's grave – Lamenting as a tool of collective grief in the era of eco-anxiety* (I8) or in individual artworks (e.g., *a Coal Peak* (I31), *Vanishing Landscape* (I11), and *Carbon Sink* (I23)). Secondly, artists respect nature by choosing wooden

Figure 4.4 Local woman plays at the *Let the Mire Play* festival, which took place in the Kauhaneva mire geopark area, Finland.

Source: Photograph by Pauliina Latvala-Harvilahti (2022).

materials to build the exhibition so that it would have a lower carbon footprint compared to industrially produced materials. Artists think it is their duty as artists to address social issues. As examples, the artwork of *Coal Peak* did not contain any human presence, but rather the human touch, and the soundscape was compiled using voices from nature. The main role was played by the mire as a carbon sink (I31). The names of the artworks further reinforce the place of mires as an environment for animals and humans, and help to identify the experiences and emotions that people may have when landscapes change or disappear. Thus, the artist of *Vanishing Landscape* does not want to provoke, but more to make a statement.

Finland has four seasons, but the climate crisis is changing the structure of the seasons. Hot summers and winters without or with only a small amount of snow are more common in large parts of the country. So, snow is being used for art as a material in Northern Finland, and ideas about climate change and changing winters have become more prominent in recent years. Reflecting on these aspects through winter art, quite a lot of material on how winters have changed is now being recorded (I29).

The Living Heritage of Mires in Art

Over the past decades and especially in recent years, it has been possible to experience art on mires in different parts of Finland. Mire art is not always intentionally artivism, but by highlighting the specificity of mire nature and its agency, also mire art, which is not based on taking a stand, can awaken people to the ecological and aesthetic value of mires and their importance as a natural and cultural heritage. Art can be a channel through which to deal with ecological concerns, and issues that are not yet publicly discussed or that are difficult to verbalise. Through mire art, people are invited to experience both nature and art in new ways. Mire art creates an affective atmosphere involving both human and non-human agencies. It emphasises the agency of nature as equal to or prior to humans. Furthermore, art may be used as a tool to ignore the human angle, and to highlight the mire as an independent actor. Thus, the mires are not just a stage for art, but also symbolic, powerful natural areas that reflect different time levels, chosen as a venue because of their rich natural and cultural heritage. It is therefore important to study how artists re-frame the mires as cultural heritage, even if they do not use this concept themselves.

By approaching the individual meanings of environmental concern and mire experiences in artists' research interviews, this chapter has pointed to new ways to get people to focus their efforts on endangered nature. The concern for the state of mires highlighted through art directs attention to the mires and their agency, making mire art also a form of place attachment. The living heritage of mires, in its socio-ecological context, is reflected both in the artists' perception of the mire as a significant theme, and as an environment for presenting art to the public. The role of the natural ecosystem has become increasingly visible in the cultural heritage of mires, and artists are not only looking at mires from a human perspective. Mire art responds to current local and global issues, and while it often includes the story of the local mire or its users, the beauty of the mire may be presented to pose ethical questions to viewers – namely: Are you ready to accept the disappearance of this nature and its uniqueness? Artists are also using digital platforms to reach audiences beyond the mires. In our understanding, positive emotions towards the mire and environmental concern can act as a driving force to act for nature, as is the case for environmental artists. Art contributes to forming the ways in which we understand mires, and other nature as well, in terms of their natural essence and autonomous agency. In so doing, art and artivism also contribute to the ways in which nature is seen as worthy of protection and preservation.

Mire artists mobilise individuals who value natural sites, underlining the importance of the growing future consciousness of both artists and the audience or volunteers. For artists, the choice to create mire art and care for positive images of mires aims to bring about a positive change in the relationship between Finns and mires. Furthermore, mire art works as a loudspeaker to local people's strong relationship to the mire, which seems sometimes to be difficult to justify in the context of land use for economic purposes. But artists have countless ways to pay attention to their homestead mires and the meaning of local mires to the people who live there. Through art, cultural meanings are often co-created, although in the current situation of the global environmental crisis and catastrophe, individual relationships and their material preconditions are under threat. However, these potential changes may be processed collaboratively in art, before such changes occur.

Mire artists represent a group whose work reflects a changing relationship with nature, and who often seek to contribute to a more sustainable understanding of nature in wider society. Our analysis has provided new insights into these soft and strong forms of expression, executed mostly by professional artists. However, there were also some self-made artists as well. What the interviewees have in common is a love of mires. Furthermore, what combines these professional and self-made artistic actions is humour and fun, but in both cases the value of mire nature is included. Artists do not have one way to express a strong climate concern. Some of the artists we interviewed had deliberately given their artworks or performances a title related to the climate crisis. But the difference between soft and strong expressions of environmental concern in artivism was also reflected in the way the interviewees talked about their other public activities. Sometimes, therefore, the strong activism of visual artists is reflected more in other areas of their lives than in their artworks. A pronounced place attachment to mire does not, however, directly imply a strong position of activism in art. So, it is acknowledged that research material on this topic cannot always be clearly categorised as soft or strong mire artivism, as its content can be more ambiguous. However, it seems that all artists are in favour of safeguarding fragile mire nature, even though they might still understand various interests and economic needs. In that sense, mire artists are cultural mediators who through their art help to identify the role of mires to communities, and who negotiate with nature on the future options we might face. Consequently, challenging people's values is not only about the art public, but also about the artists themselves.

Notes

1 For example, in Northern Finland, the area of the Viiankiaapa mire is 65.95 km^2, and most of the reserve was set up in 1988 because it was included in the national Mire Conservation Programme. Furthermore, the site belongs to the European Commission's nature programme Natura 2000.

2 Link to the web page of the research project: https://uefconnect.uef.fi/en/group/mire-trend-research-project/.

3 The acronym ITE comes from the Finnish 'self-made life'. ITE is art that is dependent on its maker's environment and experience (Knuuttila 2011, 11; ITE 2022).

4 Based on previous research, psychology researchers Robert Gifford and Andreas Nilsson (2014, 141–157) have found altogether 18 personal and social factors that influence the pro-environmental concerns and behaviors of humans: 'The personal factors include childhood experience, knowledge and education, personality and self-construal, sense of control, values, political and world views, goals, felt responsibility, cognitive biases, place attachment, age, gender and chosen activities. The social factors include religion, urban–rural differences, norms, social class, proximity to problematic environmental sites and cultural and ethnic variations' (Gifford and Nilsson 2014, abstract).

References

Campbell, Liam. 2021. *Room for the River. The Foyle River Catchment Landscape: Connecting People, Place and Nature*. Derry: Merdog.

Cervantes, Fabiola Hernandez and Maria Huhmarniemi. 2023. "Stitching for Material Sensitivity: From Traditional to Activist Embroidery." *Ruukku* 20. Taiteellisen tutkimuksen kausijulkaisu. https://www.researchcatalogue.net/view/1927154/1927929

Everyman's Rights, 8 October 2000. Environment.fi. Joint website of Finland's environmental administration. https://www.ymparisto.fi/en-us/nature/everymans_rights

Flint, Abbi and Benjamin Jennings. 2021. "The Role of Cultural Heritage in Visitor Narratives of Peatlands: Analysis of Online user-Generated Reviews from Three Peatland Sites in England." *International Journal of Heritage Studies* 28 (2): 163–179. https://doi.org/10.1080/13527258.2021.1941198

Gearey, Benjamin R. and Rosie Everett. 2021. "Running Out of Time? Peatland Rehabilitation, Archaeology, and Cultural Ecosystem Services." *Mires and Peat* 27: 1–6. https://doi.org/10.19189/MaP.2021.KHR.StA.2195

Gellatly, Kelly, Bronwyn Johnson and Guy Abrahams, ed. 2016. *Art + climate = change*. Carlton: Melbourne University Publishing.

Giblett, Rod. 2014. *Canadian Wetlands: Places and People*. Bristol: Intellect.

Gifford, Robert and Andreas Nilsson. 2014. "Personal and Social Factors that Influence Pro-Environmental Concern and Behaviour: A Review." *International Journal of Psychology* 49 (3): 141–157. https://doi.org/10.1002/ijop.12034

Golanska, Dorota and Anna Katarzyna Kronenberg. 2020. "Creative Practice for Sustainability: A New Materialist Perspective on Artist Production of Eco-Sensitive Knowledges." *International Journal of Education through Art* 16 (3): 303–318. https://doi.org/10.1386/eta_00035_1

Green, Cynthia. 2018. "Peat's Place in Art." *JSTOR Daily*. https://daily.jstor.org/peats-place-in-art/

Härkönen, Heidi, Rosa Maria Ballardini, Heidi Pietarinen and Melanie Sarantou. 2023. "Nature's Own Intellectual Creation: Copyright in Creative Expressions of Bioart." *Nuart Journal* 4 (2). Accessed 8 April 2026. https://nuartjournal.com/issues/trespass/nature-s-own-intellectual-creation-copyright-in-creative-expressions-of-bioart

Häyrynen, Liina and Sari Pynnönen. 2020. "A Review of the Concepts and Measurements for Connection to Nature and Environmentally Responsible Behaviour – a Call for Research on Human-Forest Relationships." *Current Forestry Reports* 6: 323–338. https://doi.org/10.1007/s40725-020-00131-6

Hernández, Bernardo, M. C. Hidalgo and Cristina Ruiz. 2014. *Theoretical and Methodological Aspects of Research on Place Attachment*. London and New York: Routledge.

Hyvärinen, Esko, Aino Juslén, Eija Kemppainen, Annika Uddström and Ulla-Maija Liukko, ed. 2019. *The Red List of Finnish Species* 2019. https://hdl.handle.net/10138/299501

ITE. ITE art is Finnish Contemporary Folk Art. https://msl.fi/itetaide/en/front-page/

Johansson Sevä, Ingemar and Joakim Kulin. 2019. "A Little More Action, Please: Increasing the Understanding about Citizens' Lack of Commitment to Protecting the Environment in Different National Contexts." *International Journal of Sociology* 48 (4): 314–339. https://doi.org/10.1080/00207659.2018.1515703

Joyce, Ian and Mathew Staunton. 2020. "Thinking into the Bog: Art Practice and Connection from inside a Language under Pressure." *Études irlandaises* 45 (1): 67–87. https://doi.org/10.4000/etudesirlandaises.8811

Kanerva, Anna and Ritva Mitchell. 2015. "Johdanto." In *Elävä aineeton kulttuuriperintö. Hankkeen loppuraportti*, edited by Anna Kanerva and Ritva Mitchell, 15–32. Helsinki: Cupore.

Kaukio, Virpi. 2022a. "Elämyksellinen suo." *Terra* 134 (2): 67–81. https://doi.org/10.30677/terra.112122

Kaukio, Virpi. 2022b. "Taiteilijat suon puolesta." In *Ympäristömuutos ja estetiikka*, edited by Jukka Mikkonen, Sanna Lehtinen, Kaisa Kortekallio and Noora-Helena Korpelainen, 430–469. Helsinki: Suomen Estetiikan Seura ry.

Kearney, Amanda. 2009. "Intangible Cultural Heritage: Global Awareness and Local Interest." In *Intangible Heritage*, edited by Laurajane Smith and Natsuko Akagawa, 209–225. London and New York: Routledge.

Knuuttila, Seppo. 2011. "What does ITE Express?" In *ITE Art in Finland*, edited by Seppo Knuuttila, 10–13. Helsinki: Maahenki.

Laakso, Senja and Riikka Aro, ed. 2022. *Planeetan kokoinen arki – Askelia kestävämpään politiikkaan*. Helsinki: Gaudeamus.

Laaksonen, Pekka. 2008. "Peatlands in Finnish Folklore." In *Finland – Fenland. Research and suistainable utilisation of mires and peat*, edited by Riitta Korhonen, Leila Korpela and Sakari Sarkkola, 266–271. Helsinki: Finnish Peatland Society & Maahenki Oy.

Lähde, Ville. 2022. "Arjen teoille on annettava yhteinen suunta." In *Planeetan kokoinen arki – Askelia kestävämpään politiikkaan*, edited by Senja Laakso and Riikka Aro, 82–99. Helsinki: Gaudeamus.

Latvala-Harvilahti, Pauliina. 2021. "Mire Art: Environmental Artivism and Future Agency in the Landscape of Mining." *Ethnologia Scandinavica. A Journal of Nordic Ethnology* 51: 81–101.

Latvala-Harvilahti, Pauliina. 2022a. "Kohtaamispaikkana suo: Suotaide kulttuuriperintö- ja luontoarvojen yhdistäjänä." *Terra* 134 (2): 83–97. https://doi.org/10.30677/terra.112015

Latvala-Harvilahti, Pauliina. 2022b. "Experiencing a Lament Performance in a Mire: Emotions Linked to Nature and Unsustainable Lifestyles." *Ethnologia Fennica* 50 (1): 141–167. https://doi.org/10.23991/ef.v50i1.121289

Laurén, Kirsi. 2006. *Suo – sisulla ja sydämellä. Suomalaisten suokokemukset ja -kertomukset kulttuurisen luontosuhteen ilmentäjinä.* Helsinki: Finnish Literature Society.

Laurén, Kirsi. 2008. "Mires in Finnish Arts." In *Finland – Fenland. Research and sustainable utilisation of mires and peat*, edited by Riitta Korhonen, Leila Korpela and Sakari Sarkkola, 272–277. Helsinki: Finnish Peatland Society & Maahenki Oy.

Laurén, Kirsi. 2011. "Writing the Environment: Finnish Written Nature Narratives." In *The Space of Culture – The Place of Nature in Estonia and Beyond*, edited by Tiina Peil, 107–119. Tartu: Tartu University Press.

Laurén, Kirsi. 2022. "Suo karnevalistisena tilana." *Terra* 134 (2): 99–112. https://doi.org/10.30677/terra.112016

Lehtinen, Ari. 2000. "Mires as Mirrors. Peatlands – Hybrid Landscapes of the North." *Fennia* 178 (1): 125–137.

Manzo, Lynne C. and Patrick Devine-Wright, ed. 2021. *Place Attachment. Advances in Theory, Methods and Applications.* London: Routledge.

Milton, Kay. 2002. *Loving Nature. Towards an Ecology of Emotion.* London and New York: Routledge.

Nature Conservation Act 1996. Finlex.fi. https://www.finlex.fi/en/legislation/translations/1996/eng/1096?language=eng

Nicholsen, Shierry Weber. 2002. *The Love of Nature and the End of the World: The Unspoken Dimensions of Environmental Concern.* Cambridge, MA: MIT Press.

Peatart.com: *Peatart Gallery – Treasures from the Bog.* https://peatart.com/news/2/

Pietarinen, Heidi and Amna Qureshi. 2023. "Life Between Art and Blood." *Ruukku.* https://www.researchcatalogue.net/view/1944362/1944363

Pirtola, Papu Marjatta. 2008. "Use of Peat in Textiles, Artwork and Buildings." In *Finland – Fenland. Research and Sustainable Utilisation of Mires and Peat*, edited by Riitta Korhonen, Leila Korpela and Sakari Sarkkola, 278–283. Helsinki: Finnish Peatland Society & Maahenki Oy.

Raymond, Cristopher M., Gregory Brown and Delene Weber. 2010. "The Measurement of Place Attachment: Personal, Community, and Environmental Connections." *Journal of Environmental Psychology* 30 (4): 422–434. https://doi.org/10.1016/j.jenvp.2010.08.002

Rodal, Asunción Bernárdez, Graciela Padilla Castillo and Roxana Popelka Sosai Sánchez. 2019. "From Action Art to Artivism on Instagram: Relocation and instantaneity for a new geography of protest." *Catalan Journal of Communication & Cultural Studies* 11 (1): 23–37. https://doi.org/10.1386/cjcs.11.1.23_1

Segers, Ruth, Karin Hannes, Ann Heylighen and Pieter Van der Broeck. 2021. "Exploring Embodied Place Attachment through Co-Creative Art Trajectories: The Case of Mount Murals." *Social Inclusion* 9 (4): 116–129. https://doi.org/10.17645/si.v9i4.4403

Siivonen, Katriina, Hannele Cantell, Outi Ihanainen-Rokio, Mikko Kurenlahti, Senja Laakso, Pauliina Latvala-Harvilahti, Janna Pietikäinen and Arto O. Salonen. 2024. "The Means of Achieving Sustainability Transformation." In *The Avenues to Sustainability. Key Lessons from Sustainability Science*, edited by Tarja Halonen, Kaisa Korhonen-Kurki, Jari Niemelä and Janna Pietikäinen, 215–254. Helsinki: Gaudeamus.

Sorakunnas, Esko. 2022. *The Consumer Value of Nature-based Tourism: An Examination of National Park Visitors.* Joensuu: University of Eastern Finland.

Strack, Maria, ed. 2008. "Executive Summary for Policymakers "Peatland in Global Change"." In *Peatlands and Climate Change*, edited by Maria Strack, 9–12. Jyväskylä: International Peat Society.

Suomalaisten luontosuhteet. Kyselytutkimus 2021. Helsinki: Sitra.

Takalo-Eskola, Ilkka-Juhani. 2017. *Taidan olla taiteilija.* Somero: Amanita.

Thornes, John. 2008. "A Rough Guide to Environmental Art." *Annual Review of Environment and Resources* 33: 391–411. https://doi.org/10.1146/annurev.environ.31.042605.134920

Torvinen, Juha. 2012. "Johdatus ekomusikologiaan: Musiikintutkimuksen vastuu ympäristökriisien aikakaudella." *Etnomusikologian Vuosikirja* 24: 8–34. https://doi.org/10.23985/evk.66810

UNESCO: Intangible cultural heritage domains. https://ich.unesco.org/doc/src/01857-EN.pdf

Williams, Daniel R. and Jerry J. Vaske. 2003. "The Measurement of Place Attachment: Validity and Generalizability of a Psychometric Approach." *Forest Science* 49 (6): 830–840.

Wilson, Anthony. 2005. *Shadow and Shelter: The Swamp in Southern Culture.* Jackson: University Press of Mississippi.

Interviews

Interviews conducted in Mire Trend research project (in Finnish Suotrendi-tutkimushanke). Archived in Finnish Literature Society, Helsinki, 2025.

5 A Cultural Ecology of Contemporary Encounters with Prehistoric Sites through Art

Patrick Dillon and Priscilla Trenchard

Land: We walk on it, take from it, build on it, sometimes revere it, and frequently abuse it. Ultimately, we return to it. Whereas our tenure is short, some of the material we leave behind is more enduring. Monumental structures erode gradually over centuries. Lesser artefacts are absorbed into the earth to disintegrate, which in certain conditions may take a considerable time. Ditches, banks, and earthworks of various shapes and sizes attest to ways in which we work the land. Remote sensing technology reveals traces, even if they are not visible as surface features.

Archaeologists tap into this fragmentary record. Their terms of reference, although not fixed, have a degree of prescription to reveal something of the cultural ecology of human engagement with past environments. Artists have a freer brief. They can engage with the processes of archaeology and its products, and with past and present environmental contexts. As Bailey (2014, 232) observes, when archaeologists and artists ignore the boundaries that restrict their own disciplines and work together, they can unleash the potential of pan-disciplinary thinking on major themes in the humanities and social sciences.

In this chapter, we look at the relationship between archaeology and art through the work of one of us, multidisciplinary artist Priscilla Trenchard. The artwork is focused on two sites: The Uffington White Horse, a large piece of land art dating from the Bronze Age, built approximately 3,000 years ago into a chalkland escarpment of the North Wessex Downs in Oxfordshire. It is now a national monument. The Pebblebed Cairns in the heathlands of East Devon are also associated with the Bronze Age. They were excavated in 2012 and have been re-absorbed into the landscape.

The artwork is examined within an interpretive framework developed by co-author Patrick Dillon. The framework is an extension of ideas proposed by anthropologist Julian Steward, who used the term 'cultural ecology' to explain how societies are in tension with their environments (Steward 1955). Dillon (2025) has recast Steward's ideas in the context of current thinking in neuroscience, the social sciences, and humanities, to offer a contemporary theory of cultural ecology. The reformulated framework is used here as a transactional, phenomenological way of looking at how people engage with their environments.

DOI: 10.4324/9781003727972-5

The term 'environment' is used in an inclusive sense, embracing not only the physical and biological fabric of the world, but also people's transactions with that fabric through their activities, ideas, and beliefs. The humanised environment, or cultural ecology, encompasses every facet of human behaviour, and the institutions that have developed around it, social, economic, political, religious, and moral. Cultural ecology has conceptual similarities with the idea of 'natureculture', a synthesis of nature and culture in relationships that are both biophysically and socially formed, and inseparable. Both approaches recognise that affordances arise from planetary production and recycling which are processes of renewal. Humans use affordances as resources in processes that often lead to their depletion.

Landscape is a cultural-ecological artefact which carries within it the imprint of successive generations of human engagement. Some of the surviving historical imprint may accrue special significance because of what it represents about earlier associations between people and nature. Significant parts may be preserved or represented as cultural heritage, when in addition to its historical significance, it attracts the attention of contemporary artists who make their own meaning round it.

Art, taken in a broad definition, is the application of creative skill and imagination in the production of works that are appreciated for their beauty, and more importantly in the context of this chapter, for their symbolic, emotional, and spiritual significance. Art cannot be understood as separate from the behaviour of the people who produce it, nor from the environmental context in which it is produced.

This chapter is written in the tradition of reflective practice and phenomenology, based on interacting dialogues. One of us (Priscilla Trenchard) was present at the excavation site throughout the duration of her residence at the Pebblebeds Cairns, in continuous interaction with the archaeologists. At the Uffington White Horse, her work was part of a group project with other contributing artists in on-site visits. In both cases, the artwork that arose was inexorably influenced by the experiences and dialogues. There is also a dialogue between theory and practice. Methodologically, this has involved deconstructing the artist's first-hand accounts of her style of operation, methods, and products, and exploring them as modes of engagement in the working environment.

The Framework of Interpretation

The explanation of the framework in this section is adapted from Dillon (2025). The mechanism for exploring the behaviour of people within an environmental context is the *cultural-ecological transaction*. The transaction can be examined through two dimensions that are integral to each other: the individual, *co-constitutional* moment of being, and the collective *relational* understandings that come from accumulated experience and knowledge.

Behaving within a context is a relational process, that is, it is informed by previous experiences and accumulated knowledge. The term 'relational' is derived from 'relation', meaning belonging to or characterised by, and it is this that enables distinctions to be made between behaving in one situation compared to another. The relational associations are the abstractions and conceptualisations which are the basis of how we 'make sense' of the world. But there is another dimension to the cultural-ecological transaction.

In addition to relational processes, unique, personal dimensions are simultaneously created. These dimensions are a property of the uniqueness of individual moments; they are literally constructed out of the ways in which individuals engage with the affordances of their environment as they exist in that moment: the individual, the environment, and the context all co-construct each other. These are called co-constitutional processes to distinguish them from relational processes. The term 'co-constitutional' is derived from 'constitute', meaning the whole made from its contributing parts where all the parts are actively involved in the process.

The theoretical distinction between the relational and the co-constitutional is an analytical device. In the continuous stream of moments of being which we experience as the passage of time, co-constitutional engagements immediately and seamlessly interact with relational constructs. This is co-constitutional-relational reciprocity. Surrealist Marcel Duchamp's idea of 'inframince' is a loose metaphor for it. Celia Rabinovitch (2020, 80–81) describes inframince as 'the unquantifiable point of connection'.

By definition, the co-constitutional exists only 'in the moment'; it is fleeting, but its influence can be profound. Creativity, improvisation, ingenuity, insight, and so on, typically occur 'in the moment' or in the 'flow'. The relational and co-constitutional are continually re-structuring each other in ways that are themselves relational and co-constitutional. Ference Marton (1993) explains it like this: 'systematic understanding of the world is derived through cumulative organization and rearrangement of experientially acquired understandings of the world'.

Everything we do, all cultural-ecological transactions, is based on co-constitutional-relational reciprocity. The outcomes: social norms, economic conditions, material resources, means of exchange, institutional structures, human resources, beliefs, values, attitudes, tastes, needs, wants, patterns of production and consumption, and so on, are all affordances which can resource further transactions. Archaeologists and artists, through their ideas and techniques, work with these affordances and seek to recover evidence, insights, and representations of the cultural-ecological transactions that shaped earlier societies.

We compare and contrast Priscilla's work at the two archaeological sites. We examine her modes of operation, methods, and products as cultural-ecological transactions. Our source material is the written accounts of Priscilla's work (Dillon 2016; Trenchard 2017) and the field notes, sketchbooks, and photographs produced by her at the sites.

The Artwork and Its Contexts

The environmental contexts of the two sites are archaeologically linked to the Bronze Age, but otherwise they are very different.

The Wessex chalklands of central southern England are a landscape rich with prehistoric monuments. The chalklands are now intensively farmed, but they are also places where the presence of the monuments imparts cultural and spiritual significance. The Uffington White Horse is an abstracted shape, a 'hill figure' or 'geoglyph', some 110 m long, constructed in the chalk escarpment. Given that the figure needs regular maintenance to prevent it from being overwhelmed by vegetation, its survival is remarkable. Its archaeology has been extensively researched (Miles et al. 2003; Miles 2019).

The White Horse and a large area of surrounding landscape are maintained by the National Trust as a public amenity. It is a scheduled ancient monument. The area is managed as sheep-grazed grassland. It attracts tens of thousands of visitors a year. Most of them want a day out in attractive countryside. Some people develop a spiritual connection, often choosing quiet times to engage with the Horse 'in the moment', in whatever ways that are meaningful to them. Others express their connections through art, music, and writing.

The Pebblebeds is an area of lowland heath of over one thousand hectares in the East Devon Area of Outstanding Natural Beauty. It spans several parishes and comprises a series of contiguous commons, where customary rights to graze domestic animals and collect wood for fuel were once exercised. Most of the commons are part of a private estate but they are open to the public and are highly valued by local communities and visitors. They are managed through a charitable trust, and parts of them are leased to and/or managed by the Royal Society for the Protection of Birds, the Devon Wildlife Trust, and other landed interests. The special character of the area derives from the underlying geology of Triassic Age Bunter Pebblebeds. There is a long history of occupation and use. The Iron Age hillfort at nearby Woodbury Common and numerous barrows and cairns are of significance. The archaeology, history, and contemporary land-use of the East Devon Pebblebeds are detailed by Christopher Tilley and Kate Cameron-Daum (2017).

Excavation of three mounds of pebbles known as 'cairns' was undertaken in the summer of 2009 on the Pebblebed Commons (Tilley 2017). The archaeological significance of cairns is usually associated with their use as burial sites or marker points. Priscilla was an artist in residence and worked as an integral part of the team, helping with excavation, and simultaneously producing artwork.

Priscilla's work with the White Horse was part of a group project on the theme of the monument in its landscape. The other contributing artists were Anna Dillon (www.annadillon.com), a landscape artist who paints prominent landmarks in the North Wessex Downs, and Chris Hoggett, a designer, illustrator, and artist whose first encounter with the landscape of the White Horse is recounted in his book *The Joker: Self-transformation Through Art and Dream* (Hoggett 2004). The work of the project was exhibited at the Vale and Downland Museum, Wantage, UK in 2016.

At the White Horse, Priscilla made preparatory drawings and took photographs on site, but produced the final artworks in the studio. These were mixed-media abstracts on small-to-medium-sized panels and some prints. She also produced studio work for the Pebblebeds project, but during the excavation worked in parallel with the archaeologists letting the art emerge through the experience of being in the place, utilising materials immediately to hand. The artwork produced *in situ* was on a much larger scale, and it is this work that is discussed below.

Whereas materials are typically seen as resources, within a cultural ecology they are affordances. The distinction is subtle but important, and it is a facet of co-constitutional-relational reciprocity. For example, paper is a medium on which a painting can be made; that is, it is a relational resource, it offers a surface which is different from, say, that of a canvas. But it is only at the point of being used that the particularities of the paper, its textual qualities and surface properties, that is, its affordances, influence the work and are co-constituted with it, and with the methods and skills of the artist.

At the Pebblebeds, Priscilla thought about the processes involved in making the cairns. Stones would have to be brought up from stream beds or elsewhere across the heathlands, laborious work requiring robust containers. Leather or baskets woven from willow would have withstood the hard usage. Weaving became a theme and a metaphor for life on the heathland. The dominant plants, gorse and heather, grow together, creating dense, woven undergrowth. Tracks 'weave' all over the heathland, many created by animals. Walkers, horse riders, bird watchers, and the military who train there reinforce the tracks. They become semi-permanent routes. The pebbles have some weave-like order in their placement within cairns. As each pebble layer of a cairn is unearthed, stories are woven around it.

Near the excavation site, Priscilla sourced some willow from a boggy area adjacent to a stream and started to make a basket. It was a struggle: the stems were not pliable enough. For weaving, they should be cut in the winter months and soaked. She began working on a free weaving, using plant roots collected from the excavation spoil heap, aiming to produce a large sphere strong enough to hold pebbles. She records:

> None of the roots were straight, so the shape had a life of its own, with not too much strength to the structure. As the structure grew it felt as though its weight was pushing its sides down. The inflexible and random nature of the roots and branches inevitably dictated its form. Not surprisingly it related to its surroundings. At this point, I thought I may have been getting too involved with structure, rather than working through ideas. The container could be more imaginatively conceived. The cairn is about containing, and it also covers something of importance. How could I create something that would communicate that?

Figure 5.1 *Woven Flame*, approx. 100 × 100 cm, in its heathland setting.
Source: Photograph by Priscilla Trenchard (2012).

She persisted with the idea of weaving, collecting heather and gorse, and spending hours stripping the needles off the gorse stems. She noted the repetition in the process: walking, collecting, sorting, and stripping back, a meditation of sorts. The making of the cairns would have required repetition, many journeys back and forth with pebbles, creating paths, creating stories, and remembering. Feeling despondent about the wayward nature of the weaving, Priscilla bound it with string so that further weaving might bring the whole thing together like a basket (see Figure 5.1):

Figure 5.2 The White Horse and the Manger. Drone photograph, altitude 20 m, Hedley Thorne, 2022, with permission.

> The wind took the structure, rolling it around like tumble weed, and parts broke off. One of the archaeologists suggested putting the basket on top of the cairn. Here it looked like a woven flame, and it stayed steady, singing musically and metaphorically in the wind. Its character changed depending on where it was viewed. *Woven flame,* as it became known, 'worked'. It had balance and harmony with the landscape. Gradually people took a break from excavating to look at it and take photographs. They took it seriously, emotionally.

This level of intimacy with the affordances of the environment was not possible at the Uffington White Horse (see Figure 5.2). It is managed as an historic monument. Permission is required for any activity that disrupts it physically. The only opportunity people have of immersion in the material fabric of the Horse is through periodic scourings when vegetation is cleaned away from the chalk surface. But these are managed restorations with strict rules about the materials and how they are used.

The White Horse is set in a dramatic landscape. The hill figure is both a figurative depiction of a horse and an abstraction of it. It lies close to the ancient track known as the Ridgeway, which runs past Uffington Castle, an Iron Age enclosure, and Waylands Smithy, a Neolithic chambered tomb, or long barrow. Beneath the escarpment hosting the White Horse is Dragon Hill, and beyond that a deeply ridged dry valley known as the Manger. The White Horse is possibly the earliest surviving monumental 'land art' in the

United Kingdom. The imagery has been widely interpreted: a tribal symbol, a totem, a dragon, and a 'Celtic' motive. The last of these also features on Iron Age coins and decorative metalwork. The landscape is rich in folklore and artistic and literary associations.

Not only has the archaeology of the White Horse and its landscape been extensively studied, but also for centuries the Horse has been the 'subject' of art. Early topographers and antiquarians produced illustrations, archaeologists have sketched and drawn it. Mid-twentieth-century modernist artists were intrigued by it. Eric Ravilious did a series of paintings of hill figures in the late 1930s including one depicting the Horse (Dillon 2016). Paul Nash photographed it, recognising it as a work of both design and art (Nash 1938). Several artists chose the White Horse or elements of its landscape for their contributions to the 'Landscape of the Ridgeway' commissioned by Francis Kyle for an exhibition in 1985 (Ingrams 1988). More recently the White Horse has featured on popular music album sleeves, pub signs, and pottery.

In the absence of being able to work directly with the physical fabric of the White Horse, Priscilla focused on the constant movement of light across the landscape and the transformations brought about by changing shadows. She used a sketchbook and photographs for recording field experiences. These became reference materials back in the studio where mixed-media paintings were produced by techniques of laying down and scraping off, revealing and obscuring, emphasising and toning down. The layers and textures in the artwork mirror the processes of deposition, weathering, and erosion that have shaped the landscape and the human enterprises that have fashioned the surface detail. The artworks have a multi-dimensional map-like quality (see Figure 5.3). They are abstractions in the broadest sense, never an absolute suggestion of subject, always open to interpretation.

One can create art at the White Horse site as a visitor, but there is a degree of detachment. Artistic engagement is primarily based on sensory and imaginative responses. In contrast, being an artist in residence at the Pebblebeds afforded the opportunity for immersion in both the process of excavation and the material fabric of the site as it was revealed. Working on-site offers opportunities for more nuanced work with nature's affordances. For example, the pebbles come in an array of colours and markings. Classifying them was a big part of the excavation's agenda. Each pebble was weighed, measured, washed, and categorised, its location recorded, and all this information documented. Colours were sorted too: reds/browns, white, yellow, blacks, and 'specials'. When wet, the pebbles disclose an array of colours and patterns, which probably afforded them as much attraction in the Bronze Age as now.

Priscilla adopted one of the metal grids that the archaeologists use when making plans of the layout of the pebbles. The metre square grid, with its 25 divisions, is placed over the area to be recorded and the information

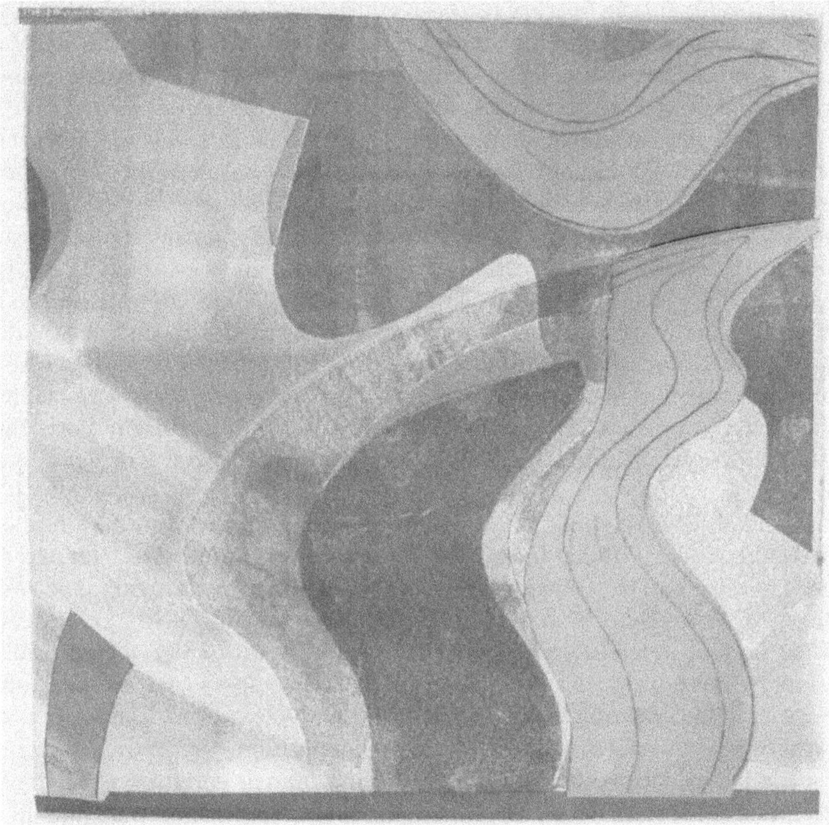

Figure 5.3 *The Way Through*. Monoprint, 25 × 25 cm, based on the landscape of the White Horse.

Source: Photograph by Priscilla Trenchard (2016).

laboriously drawn onto graph paper. She used the archaeologist's colour categories for the pebbles, but sub-divided some to give a better differentiated range: black, black with white lines, white, white quartz, grey, red, red/brown, yellow, orange, pebbles that were broken in half, and ones that looked like body parts. She arranged the pebbles in the grid (see Figure 5.4), a three-dimensional equivalent of the archaeologists' written record, using the material that was the subject of the record. This artwork was named *Pebble Grid*.

Another idea was to place pebbles on different materials and leave them in the landscape for a few weeks as a means of creating 'environmental prints'. Priscilla set up four printing experiments: (i) canvas as the print medium, placed over the pebbles on top of a cairn, and weighed down with more pebbles; (ii) watercolour paper placed on the ground near the cairn with pebbles on top; (iii) a sandwich of paper, sticks and paper with pebbles

Figure 5.4 Pebble Grid, 100 × 100 cm.
Source: Photograph by Priscilla Trenchard (2012).

on top; (iv) a sandwich of paper, grasses and sticks with paper on top and then pebbles.

After four weeks, very little had happened to the printing experiments. The heathland terrain drains quickly after rainfall and the dew dries rapidly in the ever-present wind. On her return some weeks later, after the excavation had finished, she found that paper experiments (ii), (iii), and (iv) had mainly been nibbled away leaving small scraps, but the canvas pieces had survived. They had 'prints' where the rain and soil had settled around the edges of the pebbles and left stains on the fabric. The stains included very subtle shades of greens and pinks. As the fabric had produced interesting results, Priscilla left a large piece over the whole cairn for a few months. The fabric was then covered completely with pebbles making a sort of 'sandwich' (see Figure 5.5). This was to become the *Tor Cairn Shroud*.

Over the following years, Priscilla returned to the Pebblebed cairns and noted how traces of the excavation and the remnants of her on-site artwork were gradually re-absorbed into the landscape. *Woven flame* disintegrated, leaving but a trace, detectable only by those who knew it was once there. Sometime later, at an excavation of an Early Bronze Age cremation burial on Dartmoor, a similar landscape about 60 km distant, a woven basket made from the fibrous inner bark of a lime tree turned up. The basket contained beads, wooden ear studs, and a flint flake. It was made from two woven circular discs forming a flat base and a lid, each 12 cm in diameter, joined together

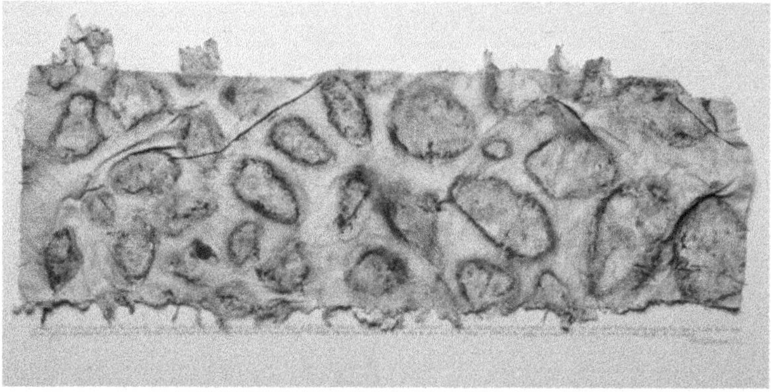

Figure 5.5 Section from *Tor Cairn Shroud*, 90 × 30 cm, mounted for exhibition.
Source: Photograph by Priscilla Trenchard (2012).

with a tube of coiled basketry. There was stitching with cow hair around the edges (Jones 2016). From the Pebblebeds Tor Cairn, on a clear day, one can look out to Dartmoor and imagine that in prehistoric times, there were craft traditions connecting artisans working across these landscapes.

The *Pebble Grid* also merged comfortably into the landscape with its colours blending into the surroundings. Plants grew through it, and its edges frayed and discomposed back into its constituent parts. This was art first emerging in the landscape and then being slowly reclaimed by it, a reverse process to archaeology which seeks to recover what the landscape reclaimed from the activities of people generations ago.

After the excavation had finished, Priscilla continued to work in her studio, using the ideas and materials from the project. She comments that these artworks, although interesting, have a sense of 'control' and formal presentation, characteristics that were not present in the landscape pieces. Priscilla is a member of the Devon Guild of Craftsmen. The Guild hosted an exhibition of her Pebblebeds work at the Riverside Gallery in Bovey Tracey, Devon in 2012. A reconstituted *Pebble Grid* and the *Tor Cairn Shroud* were the focal points of the exhibition: 'Strata: An artist's response to an archaeological project'. Later, the exhibition toured other villages in Devon.

Discussion

Priscilla Trenchard's engagement with the affordances of the two archaeological sites produced very different modes of operation and artistic outcomes. These reflect the differential interplay between the relational and the co-constitutional in the respective cultural-ecological transactions.

Much of the artwork produced today is a commodity for a consumer market. This is 'market' not only in an economic sense, but also driven by people's expectations of what art should be. Scholarly interpretation reflects this.

Typically, it is approached from either an historical perspective or through critique. The art-historical approach enables work to be placed relative to temporal, geographical, and stylistic traditions. The art-critique approach concerns the materiality and the 'voice' of the art – its production and composition, how materials and techniques are used, and what the work has to say about contemporary social and political issues. These approaches emphasise the relational aspects of art (see Daniels 2004 for a thematic overview of recent trends in relationships between landscape and art).

The White Horse in its landscape has a strong relational presence in the collective psyche. It is well defined both archeologically and artistically, the one through extensive research and excavation, the other through a body of artwork produced by many people over many generations. Together these impart an iconic status to the site. With a lineage going back several millennia, it represents continuity, a long, unbroken history of human engagement. The image of the Horse is symbolic of this heritage. It is a powerful legacy, and one from which it is difficult to break free (Dillon 2026).

Priscilla found the relational, representational presence of the White Horse overwhelming. Unlike her work at the Pebblebeds, there was no intimacy and immediacy. One could visit the White Horse, walk through the site, but never be 'in it', in the sense of being immersed in its raw affordances. There was no possibility of working with the material of the site and producing large-scale artwork on the site. And always the weight of its history and status demanded attention. She tried to overcome this by producing abstracted, mixed-media work which was influenced by the landscape but not obviously representing it. She was never entirely satisfied with the outcomes.

A paradox of working at the Pebblebeds is that the brief of working with the excavation and its materials restricted the potential range of affordances of the site, but at the same time enhanced the freedom of working with them. There was no repressive historical legacy. The co-constitutional moment of engagement was unconstrained. The artwork was integral to the story that unfolded through the excavations:

> The challenge of working in the landscape and limiting the use of materials to those at hand for creating work was both intimidating and liberating. Taking oneself out of one's comfort zone opens up unpredictable possibilities. Trying to weave with gorse and heather was impractical because of the inflexibility of the fibres. However, the resulting' basket', although not practical as a vessel, when placed into the landscape reflected its environment and took on a presence in that space. I was surprised and honoured as members of the team walked up to the cairn where the basket, known as *Woven Flame* was situated, some leaving a pebble inside as a type of offering. That was very touching.

Artistic engagement with the environment is an embodiment and expression of sensory experience co-constituted with the affordances of the environment. This has some commonality with Donna Haraway's idea of the 'Otherness-in-Connection' we have with 'Other-than-Human' living things. Harraway makes the point that we cannot 'know the Other', but we can be open to what emerges from relationships with them (Haraway 2003, 3). An extension of this view aligns with Alfred North Whitehead's organic realism, where all actual entities have a degree of sentience (Sjöstedt-H 2016, 22–23). Whereas living organisms have a self-organising unified sentience, the physical environment embodies the lesser sentiences of the molecules of which it is composed. Priscilla's artwork at the Pebblebeds was sensitive to 'nature', and to earlier human 'presence' in the landscape. At the White Horse it was a 'lesser sentience' experience.

The Pebblebeds felt special to Priscilla, 'as though it is suspended between the earth and sky, a safe, secret place'. She describes her experience as having a sacred quality to it: the repetition in some of the activities, creating a meditative ritual through weaving, the 'pilgrimage' in collecting stones, the memorial aspect of the cairns.

'Sacred' is a term that is applied to heritage sites that are deemed to be worthy of respect on account of their antiquity or cultural associations. Then, they might more appropriately be termed 'sacred places' rather than 'sites' (Dillon 2026). Sacred places come in diverse forms and are experienced and understood in different ways. For many, a sacred place is somewhere that is recognised for its spiritual significance, usually because of the way people engage with it through ceremony, worship of a deity, or acts of homage. The White Horse carries its sacredness like an emblem.

Sacred places have physical presence but 'sacredness', the spiritual dimension, is intangible. It might be generalised as a fluid accommodation between the intrinsic and extrinsic properties of place and the beliefs and values of the people who engage with it. Lucy Lippard (1983) characterises this as conditions that are amenable to developing meanings sensed but not seen. These intangible relationships are not static, each generation remakes its beliefs and values and 'sacredness' is continually re-structured and re-enacted. Here there is a developing relational tradition which sets Priscilla's work alongside the land art of people like Andy Goldsworthy, Richard Long and Anthony Gormley.

For example, Andy Goldsworthy produces work *in situ* with nature's affordances. Materials are arranged as a construction, and then left to re-enter nature through the actions of plants and animals, the vagaries of weather, and the processes of decay. Some installations are designed to disintegrate slowly, like the 'stones' on Alderney, one of the islands in the English Channel. The 'stones' are a set of large, rammed earth boulders with artefacts from the island concealed within them. In 2011, they were left in prominent places to weather. After several years, some of them had crumbled away, returning the hidden artefacts to the island. An archaeological equivalent is the experimental earthwork on Overton Down, Wiltshire. This is a large ditch and bank structure constructed in 1960 to simulate a prehistoric site. Artefacts are buried in the bank. The earthwork is excavated periodically and the state of the bank and the materials buried in it are recorded. This provides insights into the processes of decomposition and preservation in archaeological sites (Jewell and Dimbleby 1966).

Land art in its most basic form is an outcome of transactions between the artist, his or her creativity, and the affordances of the immediate environment. Archaeological excavation introduces an emergent context as material survivals of the past are gradually revealed. The artist can contribute to shaping the interpretation rather than just responding to it. Practical questions about the physicality of the site can be addressed, for example, the logistics of moving materials from one place to another, and of the sensual responses of earlier people to their environment.

The cultural ecologies of art and archaeology open some bigger questions about our long-term relationship with Earth. Across the world in caves and sheltered areas with hard surfaces there are survivals of painted or inscribed images collectively known as rock art. Some are of great antiquity, some comparatively recent. Some of it is representational, of hunting scenes or domestic activity, some abstract. But what is being represented or abstracted? The subject is not a fixed entity, rather it is a transaction between what the environment affords and the sensory response of the person who views it *at the moment of engagement*. Here is how Finnish scholar Pekka Kivikäs (2005, 8) sees it:

> The paintings live in symbiosis with the surrounding environment and are continuously changing. Rock art and rock carvings are environmental art at its most authentic. Works of rock art look different in different kinds of lighting and weather. Before they will "speak" to you, however, you must have encountered them several times, have certain basic knowledge, and be prepared to approach them without preconceptions and learned ideas about figurative art. Once a painting site and the paintings themselves touch your soul, you may form a lifelong tie and friendship with rock paintings and with the natural environment that surrounds them.

'Voices' from the past are awakened through archaeology and art. Through the physical act of investigating and representing prehistory, its practitioners are brought back to Earth, literally. They engage again with its affordances.

Some archaeological sites may accrue special significance because they are nationally or regionally important monuments, or because they have 'sacred' associations, like the Uffington White Horse. But there is a simpler way to think about sacred, one that carries a more accessible and democratic meaning. Places are sacred when people venerate the land and respect what it can and cannot afford. At the Pebblebeds, Priscilla created artwork by utilising the affordances to hand, as did the people who built the cairns some three thousand years earlier. The connectedness of people and nature is ever present. By revealing the connections, art can be a powerful advocate for the environment. Sustainability is generally thought of as living with the flows and cycles of nature. The cultural-ecological argument shifts its focus from living within the means of available resources to ensuring we do not live beyond the means of what the environment can afford.

Our ancestors lived and worked in the land through day-to-day co-constitutions with its affordances. People of the land know the land, not only as territory, but sensually: the look of it, the smell of it, the sound of it, the taste of the crops that come out of the ground. The French use the term 'terroir' for the environmental factors that give a particular character to a crop. Terroir is an expression of environmental affordances translated into sensual experiences. Understanding our roles as agents of change in cultural ecologies, large and small, provides insights into how we came to where we are now, and how we might shape a future. Archaeology can open these questions; art has the potential to take them in new directions.

This chapter is dedicated to the memory of Christopher Tilley, 1955–2024, archaeologist extraordinaire.

References

Bailey, Doug. 2014. "Art//Archaeology//Art: Letting-Go Beyond." In *Art and Archaeology: Collaborations, Conversations, Criticism*, edited by Ian Alden Russell and Andrew Cochrane, 231–250. Dordrecht: Springer.

Daniels, Stephen. 2004. "Landscape and Art." In *A Companion to Cultural Geography*, edited by James S. Duncan, Nuala C. Johnson and Richard H. Schein, 430–446. Oxford: Blackwell.

Dillon, Patrick. 2016. *In The Wake of the White Horse*. Exeter: Downwriter.

Dillon, Patrick J. 2025. *Cultural Ecologies of the Land: A Restless Dynamic of People and Place*. Abingdon: Routledge.

Dillon, Patrick J. 2026. "Sacred Places as Cultural Ecologies: The Case of the Uffington White Horse in the English Chalklands." In *Sacred Places in the Arctic and beyond: Cultural and Existential Transitions,* edited by Francis Joy, Patrick Dillon and Dawid Bunikowski. Abingdon: Routledge.

Haraway, Donna J. 2003. *The Companion Species Manifesto: Dogs, People, and Significant Otherness*. Chicago, IL: Prickly Paradigm.

Hoggett, Chris. 2004. *The Joker: Self-transformation through Art and Dream*. Cheltenham: Clover Cliff.

Ingrams, Richard. 1988. *The Ridgeway Europe's Oldest Road*. Oxford: Phaidon.

Jewell, P. and Geoffrey Dimbleby. 1966. "The Experimental Earthwork on Overton Down, Wiltshire, England: The First Four Years." *Proceedings of the Prehistoric Society* 32: 313–342.

Jones, Andy M. 2016. *Preserved in the Peat: An Extraordinary Bronze Age Burial On Whitehorse Hill, Dartmoor, and Its Wider Context*. Oxford: Oxbow.

Kivikäs, Pekka. 2005. *Kallio, maisema ja kalliomaalaus*. Jyväskylä: Minerva.

Lippard, Lucy. 1983. *Overlay: Contemporary Art and the Art of Prehistory*. New York: Pantheon.

Marton, Ference. 1993. "Our Experience of the Physical World." *Cognition and Instruction* 10: 227–237.

Miles, David. 2019. *The Land of the White Horse*. London: Thames & Hudson.
Miles, David, Simon Palmer, Gary Lock, Chris Gosden and Anne Marie Cromarty. 2003. *Uffington White Horse and Its Landscape*. Oxford: Oxford Archaeology.
Nash, Paul. 1938. "Unseen Landscapes." In *Outline, an Autobiography and Other Writings,* edited by Herbert Read, 229–231. London: Faber & Faber.
Rabinovitch, Celia. 2020. *Duchamp's Pipe: A Chess Romance*. Berkeley: North Atlantic Books.
Sjöstedt-H., Peter. 2016. "The Philosophy of Organicism." *Philosophy Now* 114: 22–23.
Steward, Julian H. 1955. *Theory of Culture Change: The Methodology of Multilinear Evolution*. Urbana: University of Illinois Press.
Tilley, Christopher. 2017. *Landscape in the Longue Durée*. London: UCL Press.
Tilley, Christopher and Kate Cameron-Daum. 2017. *An Anthropology of Landscape*. London: UCL Press.
Trenchard, Priscilla. 2017. "Woven Flame and Pebble Grid: An Artist's Interaction with Archaeology and the Heathlands." In *Landscape in the Longue Durée*, edited by Christopher Tilley, 413–427. London: UCL Press.

6 Heritage, Cultural Polylingualism, and the Ecosophical Potential of Creative Conversation in Rural Wales

Iain Biggs

This chapter draws on the work of two individuals, Lindsey Colbourne and Dr Anna Pigott. Colbourne is seen as involved in modifying and extending current notions of 'pedagogic art practices' (Bishop 2012, 274) and Pigott as involved in challenging certain basic academic presuppositions. Both work with rural heritages using creative conversation grounded in cultural polylingualism, understood here as the willingness to articulate and mediate between distinct, sometimes antagonistic, mentalities rather than languages literally understood. Both individuals' activities are informed by regional, linguistic, and cultural contexts in rural Wales. In Colbourne's case, as an artist and social activist with a background in conflict resolution, through conscious immersion in her local Welsh-speaking community and its regional environs. In Pigott's case through working as a lecturer at Swansea University responsible for the module 'Changing Ourselves and Our University', and through activity that has led her 'to think differently about art, craft, plants and bodies' (Pigott 2023, 99). While their work cannot be separated from their engagement with their place within specific environments, it has implications for others working in post-colonial rural contexts.

Colbourne and Pigott have been chosen for this exposition because their creative conversation, while specific to particular contexts, also represents emergent forms of practice. Colbourne's challenges conceptions of activist art determined by the culture industry, while Pigott's unsettles assumptions about academic research as framed by traditional notions of reason that obscure 'the reality of frames, prototypes, metaphors, and contested concepts' (Lakoff 2009, 194) that ground it.

What follows are dialogues with Pigott's text *Articulating artfulness: Exploring the ecological potential of creative conversation* (2020), while proposing a more inclusive alternative to the term 'artfulness'. It does so in order to identify the lived basis for a cultural polylingualism regarding relationships between heritage, inhabitation, mentalities, and ecologies. My aim then, given that Pigott's text is published in a specialist academic journal, is to re-inflect its implications from the more inclusive perspective of long engagement with ecologies of place (see Modeen and Biggs 2020).

DOI: 10.4324/9781003727972-6

The remainder of this introduction discusses some a priori assumptions. I then move on to discuss creative conversation/cultural polylingualism and to introduce the concept of *notitia* from which the term noticing/listening, proposed as an alternative to artfulness, is derived. In 'Two examples of creative conversation/cultural polylingualism', I discuss the relationship between the examples employed here, together with some of the issues this raises. Attention is then drawn to salient aspects of the context in which the particular work of cultural translation undertaken by Pigott and Colbourne takes place. Two further sections: Anna Pigott/Emergence and Lindsey Colbourne/*Utopias Bach* set out practical aspects of their noticing/listening within a relational context. This chapter ends by returning to a challenge Pigott makes in relation to her experience with Emergence: that of our speaking differently, from the heart – a shift entailing 'relinquishing a certain dependence on rational thought in favour of flirting with "intuition at the edge of the nonconscious"' (Pigott 2020, 885), and elaborated on here by drawing on the poet and scholar John Burnside.

Creative Conversation/Cultural Polylingualism

Creative conversation is an orientation predicated on willingness to welcome new feelings, orientations, and ideas raised by mentalities other than one's own, employing a form of open conversation in which that welcome is made explicit. It makes possible a polycultural translation between different mentalities, one able to address 'the danger of incommunicability through the protective withdrawal' (Ricoeur 1996, 4). Withdrawal that results from internalising and then taking as given a closed cultural mentality. In the present context an example would be the mentality over-reliant 'on old images of Wales' heavily polarised between the largely English-speaking and once heavily industrial areas of South Wales known as '"the valleys" and rural Wales and the Welsh language' (Evans 2022, 226). Cultural polylingualism is, then, the ability and willingness to translate and mediate, through creative conversation, between distinct and sometimes antagonistic mentalities, heritages, and other issues inseparable from daily inhabitation.

This understanding is identified with the first of Paul Ricoeur's three models 'for the integration of identity and alterity', namely 'the model of translation' (1996, 4), and with George Steiner's view that translation is central to both culture and consciousness. One that leads him to argue that, without translation, 'we would live in arrogant parishes bordered by silence' (1966, 24–25). Polylingual cultural translation thus shares Ricoeur's focus on 'a *translational ethos*' (1996, 5) and is predicated on an open and attentive noticing/listening. An attitude that, in the Humanities and Social Sciences, has often been identified as 'artfulness', 'arts of living', 'arts of socioecological transformation', 'arts of noticing', or 'arts of attentiveness' (Pigott 2020, 877), this reductively identifies it with current conceptions of art.

Two points are fundamental to the distinction made here. The first relates to presuppositions embedded in everyday language practice. Earlier I may have given the impression that I regard Colbourne as an artist and Pigott as an academic. Such conventional nomenclature is problematic. It obscures their multiple, highly nuanced and interdependent ensembles of practices; ensembles linked to more open, creative, attentive, and questioning relationships to the world's complexities. Contrary to conventional nomenclature, we are each in actuality placed in the world through a wide variety of diverse relationships that modify, complicate, and thicken out our narrative identities (Ricoeur 1994, 113–139). A more complex sense of relationality that requires that we understand ourselves as: 'always both more and less than the [conventional] categories that name and divide us' (Finn 1996, 171); the acceptance of which understanding underwrites the cultural polylingualism enacted by creative conversation. Secondly, it is necessary to note that Colbourne and Pigott consciously *place* themselves via that more fundamental relationality. Physically, in and through their attentive engagement with the complex and changing psychosocial environment of Wales, particularly in its rural aspects. Historically by their awareness of the shifting conditions of a deepening socio-environmental crisis, the result is both of increased levels of precarity (Tsing 2015) and in relation to particular collective attempts to address it.

The practice of creative conversation therefore requires a particular understanding of how participants experience themselves as consciously placed, rather than merely positioned through conforming to a given, sited, mentality. This distinction follows from the understanding that: 'if a position is a fixed posit of an established culture, a place, despite its frequently settled appearance, is an essay in experimental living within a changing culture' (Casey 1993, 31). A view that, despite differences in terminology, can be related to Doreen Massey's notion of space as the '*simultaneity* of stories so far' (2005, 9). Stories that are always implicit in the understandings of heritage and habitation.

It follows that cultural polylingualism experiences lifeworlds as dynamic rather than fixed entities. In doing so it challenges such reductive assumptions about the rural as a single 'mapped terrain with distinct boundaries' embedded in fixed and 'distinct cultural and linguistic' traditions (Rogoff 2000, 122). Instead, it recontextualises assumptions about rural lifeworlds by understanding them as taskscapes co-constituted between persons and place (Ingold 2000). As such it counters reductive conceptualisations by authorities uninvolved in such taskscapes who may view a place simply as a given position on a map, a category to be studied or enjoyed recreationally, or as the passive site of a political or environmental project. The limitations of such views becomes clear when compared to the recent, broadly ethnographic, accounts that more conversationally attend to the complexities and contradictions inherent in taskscapes (e.g., Gray 2003; Bathurst 2021; Humphries and Hopwood 1999), or that explicate the complexity of the political and historical factors that inform and contextualise them (e.g., Hayes 2020; Fowler 2020; Jamie 2008; Neal and Agyeman 2006; Shoard 1987).

Notitia

The particular form of attention that underwrites cultural polylingualism is referred to here as 'noticing/listening'. This usage is derived from *notitia* as used by the Archetypal psychologist James Hillman and distinguishes its referent from tacit cultural presuppositions implicit in terms such as artfulness. This distinction amplifies a critical point made by Pigott. Namely that, while 'the concept of artfulness' stresses associations within the arts, it can also signify 'something different, which is a perspective on creativity that is not (or not only) about processes within particular art projects, but about *a manner of being* more generally' (2020, 879). A differentiation that, more fundamentally, distances understandings of noticing/listening from the presuppositions of possessive individualism intrinsic to the rhetoric of the culture industry.

Notitia designates the noticing/listening necessary to ground 'all cultural and ecological work' (Watkins 2008, 418) and, as a life practice 'requires a de-centring from the heroic ego' and the 'activation of [...] the imaginal ego' (Watkins 2008, 421). It thus represents a fundamental shift away from possessive individualism. A shift supported by the psychologist Gemma Corradi Fiumara's call that we become: 'apprentices of listening rather than masters of discourse' so as to 'promote a different sort of coexistence among humans' (1990, 57) – a point that Hillman (2008) extends to the more-than-human world.

The psychosocial shift these distinctions indicate is particularly challenging for individuals in academia and the arts, given that both spheres are enmeshed in a *realpolitik* powerfully animated by possessive individualism. Noticing/listening as a mode of being precisely challenges that *realpolitik* by positing a life practice able to enact Ricoeur's understanding of the central role of imagination in politics, not least by questioning given and hierarchical assumptions about both knowledge and creativity.

Noticing/listening provides 'a careful attention that is sustained, patient, subtly attuned to images and metaphors' and able to track 'both hidden meanings and surface presentations' (Watkins 2008, 419). As such it is attentive to the relationality central not only to creative conversation that is 'allowed to remain open so as to await the unforeseen whilst allowing ideas to converse with time itself without any foreclosed end set out in advance' (O'Neill 2012, 12). Such attention is particularly relevant in the current socio-environmental situation in rural Wales given, for example, antagonisms between those who claim ecological authority to justify a blanket 'rewilding' of upland regions and rural communities living in those regions seeking to maintain their Welsh-speaking heritage and way of life.

Creative Conversation in a Welsh Context

Conversation as a lived practice, in the sense of a daily cultivated creative skill, has been a topic for debate for some time. In 2009 Monica Szewczyk argued that: 'to choose to have a conversation with someone is to admit them into the field where worlds are constructed. And this ultimately runs the risk of redefining not only the "other" but us as well' (2009, unpaginated). A relational understanding that moves away from the assumptions of possessive individualism towards a view of self as an 'interiorization of community' (Hillman 1994, 36); one that values 'emotional intimacy, anthropomorphic empathy, aesthetic appreciation and ritual devotion' (Hillman 2006, 335).

At the end of 2019 Colbourne decided that her future creative activity would focus on just such creative conversations with local people in the valley where she lived. Namely, in the village of Nant Peris, located between Llanberis and Capel Curig, in Gwynedd, North Wales. Conversations that would be animated by her interest in the valley's past, present, and future. She would then, critically: 'just see where that went with no outcome' in mind (Meta Morffosis 2022). That commitment served to initiate a collective 'ethico-political articulation' of an ecosophy linking 'the three ecological registers (the environment, social relations and human subjectivity)' (Guattari 2008, 19–20), both in her own practice and in the activity of the collaborative *Utopias Bach*.

Colbourne's approach builds on a rich and diverse tradition of community-engaged art practices in Wales. This tradition is indicatively identified by Iwan Bala's *Certain Welsh Artists, Custodial Aesthetics in Contemporary Welsh Art* (1999) and by *HON 2022: Women Artists in Wales* (2022), edited by Christine Kinsey. Bala, borrowing the notion of a custodial aesthetics from the indigenous Australian artist Fiona Foley, explores 'art as a form of custodianship' for a community and of a 'people's memory' (1999, 11). His focus is on artists engaged with 'the specifics of their cultural, historical, political and social heritage' so as to establish 'a voice for our identity inside and outside Wales' (1999, 15–17); an understanding that has helped provide a background for polylingual cultural conversation. Similar concerns are present in the more recent book, which is focused on the work of ten women. Its introduction stresses their commitment to 'the often conversational tone' of the 'herstories' they articulate, while engagement with place ensures that this is informed by a willingness to 'encounter connections, walk their world' (2022, 3–5).

Something of the wider tensions and opportunities animating the work presented in these two books are indicated by *Welsh [Plural]: Essays on the Future of Wales* (2022), which explores the intersectional politics of its cultural heritage, environment, and language. For example, in his essay on farming in North-West Wales, Iestyn Tyne identifies a central issue for many of those living in its rural uplands today. Namely the erosion of a way of life underwritten by an intangible heritage that, in turn, is grounded in a physically unforgiving landscape, the particular qualities of which taskscape are uniquely articulated through the Welsh language and the heritage it articulates. This region, formally occupied by 'smallholdings and plots of land which once sustained families or communities', now faces being 'rebranded as an "adventure playground" for tourists and wealthy incomers' (Tyne 2022, 144, 150). Central to his argument is the Welsh-language concept of *yr hafod a'r hendref*, an ultimately untranslatable term related to the English phrase 'hefted to the hill'. It refers to an intimate entwining of human and other-than-human being inseparable from inhabiting the particular conditions and ecology of a place as a unique taskscape. An understanding that also evokes a way of life focused on place-specific forms of 'collaboration and community' (Tyne 2022, 152).

Tyne's vision for the Welsh rural uplands is grounded in the limited achievements of the nineteenth-century Welsh land movement and the disunity between English tenant farmers and their 'Celtic peers' (Shoard 1987, 85–87), and relates to current arguments for an enlightened agriculture put forward by the biologist and science writer Colin Tudge (2016). As such it contrasts starkly with that presented by George Monbiot's *Feral* (2014). Monbiot's polemic advocates the rewilding of large areas of rural upland Wales on what he argues are strictly ecological grounds. Reading between the lines, however, his underlying concern is to create what Tyne sees as an adventure playground for extreme sports enthusiasts. Significantly, Feral ignores the intersectionality of ecological issues, history, culture, and the heritage of *yr hafod a'r hendref*.

Tyne's account of a particular form of rural life is, by contrast, predicated on a clear understanding of the politics of disunity behind the poverty in north Wales. A disunity predicated on the interweaving of land ownership and 'the complexities of class and race' (Tyne 2022, 148). An interweaving that, in terms of the understanding presented here, requires root-and-branch reassessment of assumptions about heritage, language, culture, and politics. One able to address the complexities and 'wicked questions' that dog the region's current socio-environmental situation. A situation that, however, has also generated a wide variety of cultural initiatives predicated on a point identified by Lindsey Colbourne. Namely that, contrary to assumptions implicit in Monbiot's polemic, the Welsh rural uplands are not a static cultural backwater. Rather there is a long tradition in Wales, described by Colbourne as historically a country of villages, of diverse cultural activity that is both vibrant and widely dispersed, much of it engaged with notions of place, language, and heritage.

It is against this background that creative conversation can facilitate translation between multiple viewpoints in relation to intangible local heritage, to place as taskscape, and to the different forms of affect they generate. Colbourne and Pigott are by no means alone in attempting this but have links to similar practitioners. For example, to the farmer, performance artist, and Aberystwyth University researcher Dr Ffion Jones and the artist, writer, and performance maker Dr Rowan O'Neill. All four work at grassroots level with individuals and groups in specific localities so as to facilitate psychosocial and socio-environmental change. This work adopts a 'terrestrial' approach by acknowledging the need for a sense of placed-ness in which heritage can help ground an empathetic imagination that remains open to both the past and to potential futures. An open imagination that seeks to prevent withdrawal into the opposing of, for example, the local and the global or the conflation of ecological notions of citizenship with a nationalism predicated solely on either ethnicity or language. Furthermore, they enact Latour's insistence that what is now critical to our socio-environmental relationality is: 'understanding the importance of managing to register, to maintain, to cherish a maximum number of alternative ways of belonging to the world' (Latour 2018, 15–16). In doing so they facilitate the emergence of alternative, 'affirmative imaginaries [...] founded on the intense, embodied, sensuous, connection to landscape' involving the cultivation of more, 'not less belonging' (Pitcher 2018, 169).

Creative Conversation in Action

Pigott's paper, introduced earlier, refers to Emergence as the locus of a shared and open-ended creative conversational practice that has facilitated, and remains intertwined with, a mycelial, highly diverse, community of practice advocating noticing/listening. The *Utopias Bach* collaborative is a similar community of practice, co-founded and coordinated by Colbourne. This employs a wide diversity of ways to assist in creating better places for human and other-than-human beings, particularly those on whom the current state of the world impacts most drastically. Like Colbourne and Pigott as individuals, Emergence and *Utopias Bach* seek to address a problem inherent in the creation and sustaining of a new knowledge base by those who find themselves at variance with the status quo. A problem that is central to environmental debate when located within an understanding of intersectionality. Namely, they must engage practically with the difficulty that, to the extent that official ways of producing and sustaining knowledge and culture are predicated on dominant mentalities, the seriousness and validity of all other claims will be judged with reference to those mentalities. A situation that, 'in the extreme case', results in 'nothing recognisable as knowledge', or by extension as culture, being accepted that does not conform to 'the socially dominant form' (Gibbons et al. 1994, 1–2).

While Colbourne and Pigott, like *Utopias Bach* and Emergence, both seek to address this issue, they clearly do so in different ways. Pigott's paper culturally translates the values and practices of Emergence into the language of a specialist academic journal. By contrast, Colbourne's translational work is undertaken from a point of convergence between pedagogically oriented, socially engaged, arts practices, and insights gained from her many years of professional experience in conflict resolution. This convergence also characterises, in no small part, the *Utopias Bach* collaborative she helped to establish.

Pigott's entry on Swansea University's website identifies her research as engaging with the relationships between humans and where they live, and with the historical, cultural, political, and social narratives that form and inform their perceptions and evaluations of the environment. An interest grounded in the recognition of the power of those narratives in relation to our responses to ecological crises. A form of power that is arguably under-appreciated because it is not well understood. However, as *Articulating artfulness* makes clear, her academic interests are underpinned by a number of conversational engagements that do not conform to conventional academic research presuppositions, as her 'in-depth case study of conversations facilitated by an arts-sustainability project in Wales known as Emergence' (Pigott 2020, 877) demonstrates. This study accords with Monika Szewczyk's claim that, in choosing to enter into an open creative conversation, each party risks finding themselves redefined by it. My reading of Pigott's work with Emergence is that it heightened her questioning of 'consensually framed, abstracted knowledges that merely perpetuate ecological ruination' and, in doing so, strengthened her desire 'for interacting within the ecologies of our lives in more conscious and sympathetic ways, and for appreciating the ways in which our own bodies and thoughts participate' (Pigott 2020, 886). Here Pigott extends Ruth Behar's anthropological commitment to the role of 'vulnerable observer' (Behar 1996) by engaging as an active and observant participant. An understanding further developed in a recent book chapter (Pigott 2023) and one that, as a lecturer, she seeks to pass on to her students.

Lindsey Colbourne (*Heledd Wen*) engages in creative conversation in a variety of contexts and with multiple and diverse conversants. Her creative activities include those oriented by place-based artistic concerns, along with a variety of engagements related to conflict resolution, public participation, collaboration, and training. Her involvement in, and co-ordinating of, the experimental collaborative *Utopias Bach – Revolution in Miniature*, provides a focal forum for her combined interests. Much of her work is entangled in her immediate locality and the surrounding region seen as a taskscape. The extensive interview referenced here, together with the material on her website, evidences her primary commitment to creative, culturally polylingual, noticing/listening. This commitment is inseparable from her literal bilingualism. Both commitments reflect a desire to understanding and translating between the assumptions, explicit, or tacit, embedded in the particular forms of expression and vocabulary used to articulate different, sometimes antagonistic, orientations, and modes of inhabitation. Those that might be identified with categories like 'professional', 'vernacular', 'local', 'cosmopolitan', 'artistic', 'scientific', 'Welsh', 'English', and so on.

Framing the Two Individuals' Practices

Pigott and Colbourne's cultural translation is framed by concerns similar to those identified by English sociologist Ben Pitcher's discussion of the intersection of race, nationalism, and landscape belonging. Pitcher argues that relations with the earth, with place, and with iconographies of lived experience are necessary to our understanding of citizenship and its relationship to heritage, given that these underpin rural belonging. He adds that any attempt to create the connections necessary for an appropriate response to environmental change depends, in turn, on the construction of modified or alternative 'affective economies of citizenship' (Pitcher 2018, 164). Affective economies in which cultural activity is central to furthering an inclusive inter-meshing of psychic, social, and ecological economies of care.

Pitcher's argument pivots on his reading of events at Stonehenge and, as such, has a particular, and specifically English, context. As such it does not address the relationship between language, culture, and landscape that must inflect discussions of Welsh heritage, identity, and belonging. However, his examination of the obstacles created by fixed, affectively generated presuppositions about primordial ethnicity allows him to advocate alternative practices and discursive frameworks relevant to giving different meanings to the experience of the earth, place, and iconographies of lived experience in rural Wales. Nonetheless the practical question of how best to develop affective economies of ecosophical citizenship there remains? More specifically, what kinds of persons and practices might bring an ecosophically based caring across diverse mentalities into play? This, after all, is the change necessary to address the fraught and complex relationships between the cultural and linguistic heritage of Welsh rural and upland communities and, for example, advocacy for large-scale blanket 'rewilding'?

As already indicated, the two examples of cultural polylingualism offered here extend and elucidate existing alternative practices and their discursive framings. While they acknowledge and respect the affect grounded in monolithic assumptions of a 'primordial ethnicity' associated with the Welsh language and consequent notions of place and heritage, they also seek to re-frame it. Through engaging with questions of language, affect, and heritage in particular taskscapes, they direct their advocacy towards enabling a cultural flexibility and porosity. Qualities opened up by creative practices that privilege an extended noticing of/listening to a wide range of relevant differences and similarities. Consequently, and to reiterate the reason for my choice of the term noticing/listening over artfulness, Pigott's engagement with Emergence and Colbourne's with *Utopias Bach* are best understood as related to creative practices seen as 'no longer only art'. Practices in which various forms of art-related methods 'are recapitulated, ooze out and become feral in combination with other forms of life' so as to function as 'cultural entities, embodied in speech, texts, sounds, behaviours and the modes of connection' (Fuller 2011, 45). Practices that, by furthering polylinguistic cultural translation, can generate alternative affective experiences. The resulting recalibration of experiences can allow them to be re-oriented towards broader, more inclusive, more ecologically informed understandings and actions within the wider framework of intersectionality. This is the imaginative practice at the heart of the conversational work of Pigott and Colbourne, of Emergence and *Utopias Bach* and, more recently, of Pigott's involvement in *Climate Lab*, set up by Swansea University. (See references for a link to *Climate Lab*.)

Anna Pigott/Emergence

Conventionally identified as an academic specialist in Human Geography, Environmental Dynamics, and Climate Change, Pigott currently teaches a module on the Geographies of Climate Action and Activism. A role indicative of concerns at the intersections of cultural geography, ecology, climate, and activism that question the normative expectations of the Environmental Humanities as an academic discipline. This is relevant here for two reasons. Firstly, *Articulating Artfulness*, while of necessity conforming to the requirements of academic journal publication, encourages its readership to give greater credence to particular experiential approaches that acknowledge a conversational orientation. An orientation that challenges both normative disciplinary presuppositions and conventions concerning hierarchies of knowledge-based authority within academia. A desire carried over into her participation in the cross-disciplinary Welsh project *Narrating Rural Change*. This project was concerned with the re-imaging and narrating of rurality so as to evaluate what is 'being lost and gained through rurality's social and territorial transformations today' (Stringer 2018, v). Secondly, because writing the paper is itself the realisation of a psychosocial translation between the concerns and assumptions of what, conceptually speaking, are two quite distinct mentalities. On the one hand that of a specialist academic readership that identifies its status with the authority of the British Institute of Geographers. On the other, the mentality enacted by Emergence as a Welsh open-ended art, facilitation, and ritual practice project seeking to embody values of creativity, care, and compassion by nurturing and creating spaces for dialogue, inspiration, and change. The paper is, then, an example of the articulation of a cultural polylingualism as creative translation between those mentalities.

As noted above, what appears in Pigott's account of her engagement with Emergence is less the intellectual research concerns of a participant observer than a felt need resulting from her being an active and observant participant in Emergence's activities. Consequently, her paper also enacts the need to address an underlying tension concerning creative conversational practices in relation to education. One in which its 'transformative power' has the potential to elucidate and enact ways in which creative energy can be fostered 'in service of personal and ecological transformation' (Pigott 2020, 880). While 'a sense of skill, grace, creativity, and improvisation (among many other qualities)' are as useful in an educational context as in 'everyday life on a planet in crisis' (Pigott 2020, 880), such qualities are largely treated as tangential by those managerially tasked with ensuring universities work within the terms of the audit culture created by the British Government's Research Excellence Framework. Terms on which a university's survival is increasingly dependent. What is at stake here, educationally, is indicated by Pigott's observation that the conversations in which she participated at Emergence events: 'never had an explicit aim or agenda, other than a broadly stated intention to "hold space" for whatever thoughts and feelings arose in response to the theme of the event' (Pigott 2020, 881). This openness echoes Colbourne's aim quoted earlier. It was precisely this conversational orientation that Pigott helped bring to the *Narrating Rural Change* project and that, in the context of tertiary education, suggests the radicalism of her paper.

Lindsey Colbourne/*Utopias Bach*

Lindsey Colbourne's work offers an exemplary model of engagement with issues of heritage and environment present in inhabitation. One enacted via an ensemble of practices interwoven with collaborations with diverse others, whether her immediate neighbours, human and other-than-human, or similarly oriented practitioners and participants in the experimental collaborative *Utopias Bach* (web address in references). In short, she holds 'an alternative view of creative activity in which art acts to animate ensembles of heterogeneous skills and concerns, facilitating in turn processes of "mutual accompaniment" necessary to enact a geopolitics of the terrestrial' (Biggs 2021, 269).

Colbourne moved to Wales in 2012, learnt the Welsh language, and began a wide-ranging engagement with a spectrum of contemporary arts practices informed by her degree in psychology and 20 years of international experience as a pioneer of conflict resolution and public and stakeholder participation. A number of Colbourne's ongoing projects are of particular relevance to reflection on the intersection of heritage, language, and environmental concern. The *Sgwrs Dyffryn Peris Conversation* (web address in references), begun in 2020, is a multifaceted experiment that brings together her background in facilitation, concern with socio-environmental issues, and her specific commitment to the Dyffryn Peris locality. This project is primarily concerned with one-to-one conversations that are open to finding possible common ground through exploring both the physical ('natural') and intangible heritage of her home valley. Her aim is to discover 'what happens in a place – a place in transition, as all places are – when we explore that place using the basic building block of conversation and connection' (Biggs 2021, 269). It is indicative that her digital record of this ongoing work includes, among its six main headings: 'Conversations', 'A Space for Things that Emerge', and 'Links'.

Much like the traditional Russian doll, each of her headings has nested within it numerous inter-related conversations and sub-projects. The sub-headings of different aspects of her work (in Welsh and with the English titles given here) include the following. 'Plants', which references the arboreal history of the valley. 'Birds', which documents personal and ecological activity following an encounter with two cuckoos in the valley; 'Herstory', which narrates the lives and deaths over time in a quarry household, how women's lives were hidden, the slow shift from Welsh to English, women's role in the continuity and longevity of the household, including connections with art and literature, and the 'interconnectedness of lives between past and present households in Nantperis'. Along with 'Water', 'Nantperis citizenship', 'Rock', 'Growing', 'Merched Chwarel' (a collaborative work connected to the quarries local to where the four collaborators involved live and work), 'Digging', and the 'Radical Art of Conversation'. A list that gives some sense of the scope and inclusiveness of her noticing/listening in relation to an exploration of the human and other-than-human taskscape and its heritage in which she finds herself placed, 'a simultaneity of stories so far' in Massey's sense.

Colbourne's approach throughout her work is predicated on creating the types of conversational spaces that allow humans and other-than-humans to meet in order 'to explore something together'; a view based 'on the idea that in order to really get a robust grasp on something you need to have as many perspectives as possible and bring them together to … understand the … deeper common ground' (Biggs 2021, 269). This then is a practice that, as she insists, is characterised by emphasis on starting from 'a point of not knowing' or of what cannot be articulated; one that allows for the holding of a space 'where you can hear and learn from and change in, in the space that you create between people and nonhumans' (Meta Morffosis 2022). Here Colbourne's practice echoes that of Emergence as reported by Pigott. It also amplifies Ben Pitcher's argument that more accurate perceptions of landscape involve 'an unsettling, an estrangement, and a "kind of wilful unknowing"', that undermine the notion of landscape as a fixed and given heritage informing 'experiences of belonging uninflected by the distancing dynamics of modernity and urbanity' (Pitcher 2018, 168). In short, Colbourne's engagement with Nantperis and the surrounding region undertakes this 'unknowing' as an a priori of practice that, in its generation of affects, seeks to open up the possibility of alternative forms of belonging and solidarity. It does so by gesturing towards the types of alternative frameworks and different temporalities that Pitcher suggests might unpick 'British' (more strictly English) nationalist appropriations of Stonehenge (Pitcher 2018, 168).

Colbourne's work has both informed, and been informed and animated by, the *Utopias Bach* collaborative, which she both helped co-found and for which she acts as coordinator. This is an ongoing, co-created, socially engaged grouping in which regular creative conversation acts as a catalyst for fostering experiments predicated on multiple forms of conversational exchange. The collaborative's development has been a meandering organic process that, drawing on an image proposed by artist and member Lisa Hudson, is captured by the growth pattern of a strawberry plant. Its various 'runners' have included interventions and experiments. These include *Merchad y Tir*, which creatively explores the relationship between women – particularly those speaking Welsh and of colour – and land, with a view to activating transformations involving environmental and social justice, and exploring how Normal Site Campus at Bangor University can embody a *Manifesto of Inter-Species Kindness*. A project undertaken jointly with The Hwb Bangor programme to explore techniques of 'multi-modal embodiment for kinship in kindness, emerging and in fluid connectivity with place' and working towards 'a Manifesto beyond the text' (*Utopias Bach* -website, 12 November 2022). It is no more possible to predict the collaborative's future activity than it is to pre-determine the extent of a particular strawberry plant or mycelial network, or to anticipate the conclusion of an open conversation.

In Conclusion: Creative Conversation as Poiesis

The creative conversations concerning life in rural Wales initiated by Anna Pigott/Emergence and Lindsey Colbourne/*Utopias Bach* continue to play out in their specific localities, but they also resonate more widely. The larger challenge they propose is suggested by a notion offered in Pigott's account of her experience with Emergence; namely the need to speak differently, from the heart. A challenge that requires a reader, and particularly those who read from a critical analytic perspective, to relinquish 'a certain dependence on rational thought in favour of flirting with "intuition at the edge of the non-conscious"'. A re-orientation proposed with a view to the reader gaining access to intuition as 'a rigorous process that agitates at the very limits of an encounter with the as-yet-unthought' (Pigott 2020, 885). One consistent both with Hillman's conception of *notitia* as a life practice requiring both the de-centring of the heroic ego and the activation of an imaginal ego, as it is with Gemma Corradi Fiumara's concerns referenced earlier.

While in many respects this chapter's concerns accord with points made by Pigott in relation to Emergence's work in Wales, there is also a degree of divergence. I have suggested that we place less emphasis on the link, implicit in the term 'artfulness', between creativity and the arts as currently conceived, and place more on a wider and older conception of creativity. Namely, in a conversational poiesis arising 'through daily activities of tending' (Pigott 2020, 887) predicated on noticing/listening. An orientation consistent with John Burnside's view that *poiesis* 'includes not only the celebrated imaginative achievements of artists and poets but also the ways in which we all make our world habitable from day to day, in the flow of the dailiness of life' (2021, 82). As the activities of Pigott and Colbourne suggest, creative work still associated with the category 'art', where it has now 'gone feral', can serve as an important catalyst for reconsiderations and re-positionings of the relationship between heritage, inhabitation, mentalities, and ecologies, both in rural Wales and elsewhere, through acts of creative conversation that enable cultural polylingualism.

To conclude, this chapter seeks to show that we have much to gain from adopting noticing/listening as a basis for creative conversation as a mode of translation between distinct, even antagonistic, mentalities. As with the forms of relationality that link Pigott with Emergence and Colbourne with *Utopias Bach*, such a life practice should allow us to open to, and act on, new, more inclusive and complex approaches to articulating relationships between heritage, inhabitation, mentalities, and ecologies.

Acknowledgements

I would like to express my gratitude to all those involved in the various projects referred to here, and in particular to Anna Pigott and Lindsey Colbourne.

Bibliography

Bala, Iwan. 1999. *Certain Welsh Artists: Custodial Aesthetics in Contemporary Welsh Art*. Bridgend: Seren.

Bathurst, Bella. 2021. *Field Work: What Land does to People & what People do to Land*. London: Profile.

Behar, Ruth. 1996. *The Vulnerable Observer: Anthropology that Breaks Your Heart*. Boston, MA: Beacon.

Biggs, Iain. 2015. '"Incorrigibly Plural"? Rural Lifeworlds between Concept and Experience.' *Canadian Journal of Irish Studies* 38 (1–2): 260–279.

Biggs, Iain. 2021. "Ensemble Practices." In *The Routledge Companion to Art in the Public Realm*, edited by Cameron Cartiere and Leon Tan, 269–278. London: Routledge.

Bishop, Claire. 2012. *Artificial Hells: Participatory Art and the Politics of Spectatorship*. London: Verso.

Burnside, John. 2021. *The Music of Time: Poetry in the Twentieth Century*. London: Profile.

Casey, Edward S. 1993. *Getting Back Into Place: Towards a Renewed Understanding of the Place–World*. Bloomington: Indiana University Press.

Chetty, Darren, Hanan Issa, Grug Muse and Iestyn Tyne. 2022. *Welsh (Plural): Essays on the Future of Wales*. London: Repeater.

Climate Lab. 1 October 2023. https://emergenceuk.blogspot.com/

Colbourne, Lindsey. 2020. *Sgwrs Dyffryn Peris Conversation*. https://lindseycolbourne.com/prosiectiauprojects#/sgwrs-dyffryn-peris-conversation

Colbourne, Lindsey and Sarah Pogoda. 2022. Interview with Lindsey Colbourne. https://www.youtube.com/watch?v=FzkjFbhsIIY Emergence. https://emergence-uk.org/

Evans, Dan. 2022. "Reconstructing Welshness – Again." In *Welsh (Plural): Essays on the Future of Wales*, edited by Darren Chetty, Hanan Issa, Grug Muse and Iestyn Tyne, 214–119. London: Repeater.

Finn, Geraldine. 1996. *Why Althusser Killed His Wife: Essays on Discourse and Violence*. Atlantic Highlands: Humanities Press International.

Fiumara, Gemma Corradi. 1990. *The Other Side of Language: A philosophy of Listening*. London: Routledge.

Fowler, Corinne. 2020. *Green Unpleasant Land: Creative Responses to Rural Britain's Colonial Connections*. Leeds: Peepal Tree.

Fuller, Matthew. 2011. "Art Methodologies in Media Ecology." In *Deleuze, Guattari and the Production of the New*, edited by Simon O'Sullivan and Stephen Zepke, 45–55. London: Continuum.

Gibbons, Michael, Camille Limoges, Helga Nowotny, Simon Schwartzman, Peter Scott and Martin Trow. 1994. *The New Production of Knowledge: The Dynamics of Science and Research in Contemporary Societies*. London: Sage.

Gray, John. 2003. "Iconic Images: Landscape and History in the Local Poetry of the Scottish Borders." In *Landscape, Memory and History: Anthropological Perspectives*, edited by Pamela Stewart and Andrew Strathern, 16–46. London: Pluto.

Guattari, Felix. 2008. *The Three Ecologies*. Translated Ian Pindar and Paul Sutton. London: Continuum.

Hayes, Nick. 2020. *The Book of Trespass: Crossing the Lines that Divide Us*. London: Bloomsbury.

Hillman, James. 1994. '"Man Is by Nature a Political Animal" Or: Patient as Citizen.' In *Speculations after Freud: Psychoanalysis, Philosophy and Cult*, edited by Sonu Shamdasani and Michael Munchow, 27–40. London: Routledge.

Hillman, James. 2006. "Ten Core Ideas." In *James Hillman: City and Soul*, edited by Robert J. Leaver, 331–336. Putnam, CT: Spring.

Hillman, James. 2008. *James Hillman: Animal Presences*. Putnam, CT: Spring Publications.

Humphries, Steve and Beverley Hopwood. 1999. *Green and Pleasant Land: The Untold Story of Country Life in Twentieth Century Britain*. Basingstoke: Channel 4.

Ingold, Tim (2000) *The Perception of the Environment: Essays in Livelihood, Dwelling and Skill*. London & New York: Routledge.

Jamie, Kathleen. 2008. "A Lone Enraptured Male." *London Review of Books* 30 (5), 6 March 2008: 25–27.

Kinsey, Christine, ed. 2022. *HON: Women Artists in Wales*. Swansea: H'mm Foundation.

Lakoff, George. 2009. *The Political Mind*. London: Penguin.

Latour, Bruno. 2018. *Down to Earth: Politics in the New Climate Regime*. Cambridge, MA: Polity Press.

Massey, Doreen. 2005. *For Space*. London: Sage.

Meta Morffosis 2022. Interview between Lindsey Colbourne & Sarah Pogoda. Youtube: https://www.youtube.com/watch?v=FzkjFbhsIIY

Modeen, Mary and Iain Biggs. 2020. *Creative Engagements with Ecologies of Place: Geopoetics, Deep Mapping and Slow Residences*. London: Routledge.

Monbiot, George. 2014. *Feral: Rewilding the Land, Sea and Human Life*. London: Penguin Books. Narrating Rural Change. https://ruralchange.swansea.ac.uk/

Neal, Sarah and Julian Agyeman. 2006. *The New Countryside? Ethnicity, Nation and Exclusion in Contemporary Rural Britain*. Bristol: The Policy Press.

O'Neill, Paul. 2012. "To Go Beyond – The Emergence of the Durational Commons." In *Surface tension Supplement 5: Beyond Utopia*, edited by Sophie Warren and Jonathan Mosely, 10–15. Berlin: Errant Bodies Press / Surface Tension Supplement 5.

Pigott, Anna. 2020. "Articulating Artfulness: Exploring the Ecological Potential of Creative Conversation." *Transactions of the Institute of British Geographers* 45 (4): 877–890.

Pigott, Anna. 2023. "Willow, Weaving, Worlding and a Politics of Change." In *Art and Creativity in an Era of Ecocide*, edited by Anna Pigott, Owain Jones and Ben Parry, 99–115. London, New York & Dublin: Bloomsbury.

Pitcher, Ben. 2018. "Race, Nationalism and Landscape Belonging: Stonehenge on the Summer Solstice." In *Rurality Re-imagined: Villages, Farmers, Wanderers, Wild Things*, edited by Ben Stringer, 160–171. San Francisco: Applied Research + Design Publishing.
Ricoeur, Paul. 1994. *Oneself as Another*. Chicago, IL: University of Chicago Press.
Ricoeur, Paul. 1996. "Reflections on a New Ethos for Europe." In *Paul Ricoeur: The Hermeneutics of Action*, edited by Richard M. Kearney, 3–13. London: Sage.
Rogoff, Irit. 2000. *Terra Infirma: Geography's Visual Culture*. London: Routledge.
Shoard, Marion. 1987. *This Land is Our Land: The Struggle for Britain's Countryside*. London: Paladin.
Steiner, George, ed. 1966. *The Penguin Book of Modern Verse Translation*. Harmondsworth: Penguin.
Stringer, Ben, ed. 2018. *Rurality re-imagined: Villages, Farmers, Wanderers, Wild Things*. Applied Research + Design Publishing.
Szewczyk, Monika. 2009. "The Art of Conversation Part 1." e-flux journal Issue 3, February 2009. https://www.e-flux.com/journal/03/68546/art-of-conversation-part-i/
Tsing, Anna L. 2015. *The Mushroom at the End of the World: On the Possibility of Life in Capitalist ruins*. Princeton, NJ: Princeton University Press.
Tudge, Colin. 2016. *Six Steps Back to the Land: Why We Need Small Mixed Farms and Millions More Farmers*. Cambridge: Green Books.
Tyne, Iestyn (2022) 'A Clinging Existence: Farming in the Uplands of North-West Wales' in Chetty, Darren, Muse, Grug, Issa, Hanna Tyne, Iestyn *Welsh [Plural]: Essays on the Future of Wales*. London, Repeater Books.
Utopias Bach. https://www.utopiasbach.org
Watkins, Mary. 2008. '"Breaking the Vessels": Archetypal Psychology and the Restoration of Culture, Community, and Ecology.' In *Archetypal Psychologies: Reflections in Honor of James Hillman*, edited by Stanton Marlan, 414–437. New Orleans, LA: Spring Journal.

7 Mining Traditional Cosmology

Dúchas as an Alternative Nature–Culture Future

Mairéad Nic Craith

This chapter explores the Gaelic concept *dúchas* and its potential for envisioning an alternative human relationship with nature in local places. The inquiry is set within the theoretical framework of 'Patchy Anthropocene', which emphasises activism at a local rather than at a global scale. Drawing on *dúchas* as an indigenous[1] concept, the contribution examines the pre-colonial Irish relationship with nature and those elements that were diluted or lost during the process of Anglicisation. With reference to Jonathan Lear's (2008) *Radical Hope*, this chapter argues that a re-visiting of a *dúchas*-rooted ontological framework would facilitate an alternative human relationship with nature, and one that would promote sustainable behaviour in local places. This chapter is written from the perspective of an Irish migrant in Scotland who grew up with Irish-Gaelic influences in the household.

Patchy Anthropocene

We are living in a time commonly referred to as Anthropocene (see Chapter 1, in this volume). Its popular use is credited to chemist Paul Crutzen and diatom researcher Eugene Stoermer, who argue that we are living in a new geological era – one that has been heavily influenced by *Homo sapiens* (Crutzen and Stoermer 2000). Anthropocene is based on the premise that humans are the single most influential species on the planet. It hypothesises that the significant climate change currently experienced is rooted in human behaviour, which has fundamentally altered biological systems on the planet. In recent decades, there has been a substantial increase in destructive behaviour that has led to global warming, carbon dioxide emissions, and ocean toxification. The great acceleration of destructive human activity has been linked with many factors, including the industrial age, the rise of urban living, and the atomic bombs (McNeill and Engelke 2016). Globalisation and technological innovation have also been linked with the rise of toxic activity on the planet. The destruction of the planet has ushered in a new era of alienation of humans from the earth.

The Anthropocene concept is increasingly regarded as controversial. One of the criticisms levelled against it is the power it credits to humans. Tsing

DOI: 10.4324/9781003727972-7

et al. (2017) propose that the concept of *anthropos* has become so overwhelming that its force could destroy the earth. A further criticism directed against the concept is the implied homogeneity of destruction across the planet. Donna Haraway (2015, 12) suggests that the concept lacks nuance as it implies that all of *Homo sapiens* is equally responsible, rather than those in the global north in particular. Moreover, it dilutes the role of the male human species in polluting the earth. 'One might consider the Anthropocene to be a kind of "Men-o-scene"' (Bateson 2021b, 56). Many feminist academics argue that male human beings have not only dominated women and slave peoples, but they have also disregarded nature (see Adams and Gruen 2013; d'Eaubonne 2022; Gaard 1993).

Rather than rejecting the concept outright, some academics have adapted it. 'Patchy Anthropocene' is designed to counter the impression of a homogeneous rate of destruction across the planet. Tsing et al. (2019) propose that the Great Acceleration is best grasped through engagement with many small and local rhythms. This is modification rather than a rejection of the concept of 'Anthropocene'. The big story of destruction is re-envisaged as a culmination of seemingly minor events. This approach situates climate change in the local and zooms in on particular 'patches' of landscapes. Tsing et al. (2019, 186) explain that they are 'attending to specificity without being parochial'. They examine landscapes in terms of networks of activity, both human and non-human. They refer to these sites as 'beings in landscapes', which takes into account the diversity of multispecies in local places. It neither imagines a homogeneous human race nor an evenness of consequences from factors such as the industrial revolution or the Capitalocene. This intersectional, anthropological approach topples the primacy given to human behaviour in the concept of Anthropocene. Patchy Anthropocene engages with multispecies relations to explore networks of activity and their ecological consequences in specific locations.

In common with Patchy Anthropocene, folklore tends to focus on the local. Woven together over several centuries, the human and non-human narratives are linked with distinctive places and their geological activity. Bateson (2021a, 119–120) argues that folklore 'tends to be grounded in the hyper-local, embodying intergenerational notions of place and community'. This chapter proposes that a re-engagement with traditional folklore can teach us lessons for sustainable futures. It places folklore on the same spectrum as science – but at the opposite end. In *Braiding Sweetgrass*, Robin Wall Kimmerer (2020) showed the relevance of traditional wisdom as a complementary approach to Western scientific knowledge.

My focus on the Gaelic concept of *dúchas* is designed to explore relationships with the land and engagement with non-human species in specific places (Nic Craith 2025). This contribution considers the relationship between these traditional folk-narratives and the indigenous language of Ireland. It queries whether the loss of an indigenous language in Ireland through the process of colonisation impacts on the relationship with nature. Examining the vulnerability of the Irish context, it asks whether extinction is inevitable or whether an alternative approach, with an emphasis on local places, would promote a renewal of sustainability.

Dúchas

The idea of *Dúchas* (Irish-Gaelic), or *Dùthchas* (Scottish-Gaelic), has attracted much attention in recent years, particularly in Scotland (McFayden and Sandilands 2021; Meighan 2022; Ní Mhathúna 2021; Nic Craith 2023). It is a culture-specific concept that does not translate easily into English. In part this is because its meaning is complex. It can be explained as an ontological or conceptual tool that has been passed on in the oral, Gaelic tradition from one generation to the next. MacKinnon and Brennan (2012, 9) define it as follows: 'Dùthchas is a word of the land – it is derived from the Gaelic word "dú/dùth" which can mean "earth" or "land" – [with] the emotional energy of belonging and responsibility the word conveys'. There is an intimate connection between the language and the land. As Chiblow and Meighan (2022, 209) assert, 'Language is land and land is language'. Wierzbicka (1997, 2) describes *dúchas* as 'a strong sense of place that is both ancestral as well as contemporary. It perpetuates Gaelic ecological ideals'. Newton (2019) explains how Gaelic words such as *Dùthchas* encode, transmit, and reinforce particular ways of thinking about the relationship between people and nature.

For the purposes of this chapter, I would like to analyse *dúchas* in terms of its differing attributes – all of which have relevance for future relationships with nature. *Dúchas* is primarily land-based and treats landscape as a living rather than an inanimate object. It is quite local in that meanings are attached to specific areas that can be quite small. Like sustainability, it is a relational concept that treats humans, animals, insects, land, and water in relation to one another. For this reason, it is multispecies in outlook and does not place humans at the top of the evolutionary pyramid. *Dúchas* may be represented diagrammatically as outlined in Figure 7.1.

Dúchas can be described as an indigenous concept in that it emerged from the Gaelic culture, language, and landscape. The British cartographer Tim Robinson (2003, 12) described the Irish language as having grown 'like ferns out of the clefts of the rock'. Elsewhere he describes Irish as 'an emanation of the land of Ireland, of this segment of the earth's surface and its moody skies' (Robinson 2007, 22). A sense of *dúchas* is best captured in the Gaelic proverb, *Ag teacht abhaile ar ár ndúthchas* (coming into our own, to where we belong; becoming what we were). In using the term 'indigenous', I am referring to the ontology or paradigm that existed in Gaelic, pre-colonial Ireland and Scotland, both of which suffered under the expansion of the British Empire. While the process of colonialisation is not generally associated with Western Europe, Hechter (1975) argued that some Western

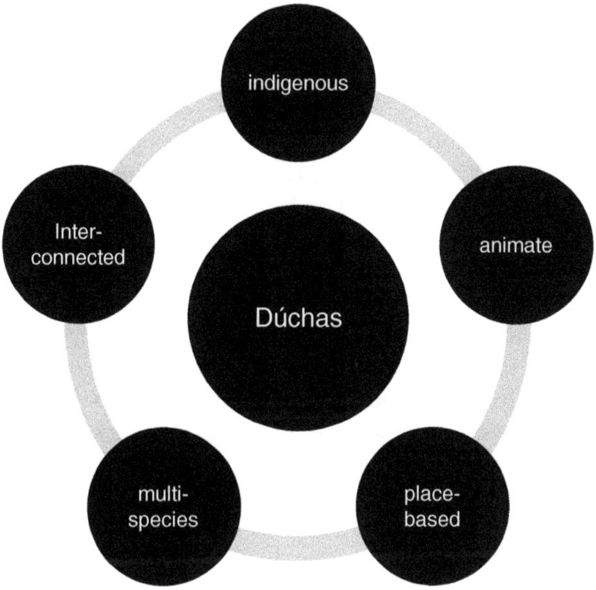

Figure 7.1 Attributes of Dúchas. Figure by Mairéad Nic Craith.

European countries suffered a form of internal colonisation during a process of state-building when core states expanded their control to include culturally distinct peripheries such as Scotland. Iain Crichton Smith (1982) uses the concept of historical 'interior colonisation' to describe the Anglicisation of Gaels in Scotland, which left them devoid of their 'webs of significance' (Geertz 1973). I am describing a process that prompted 'symptoms of human-ecological disconnect, alienation and loss of meaning – an indicator of just how far our human psyche and culture has become divorced from our natural environments' (McFadyen and Sandilands 2021, 163).

Interconnectedness

A key feature of *dúchas* is interconnectedness. In the Gaelic world, the barriers between humans and non-humans were less stringent (Nic Craith 2023). There was an ecological balance between differing entities. As Meighan (2022, 5) explains, 'the importance of this concept of the connectedness and inter-relationships between land, people and culture, held in the word "*Dúthchas*", cannot be overestimated'. It captures the way that land, people and community, past and present, are interlinked – connections which are maintained by practices, language, and other means. Keith Basso (1996, 54) points to a similar cosmology among Native Americans when he notes that 'dwelling is said to consist in the multiple "lived relationships" that people maintain with places.' Deborah Bird Rose (2013, 137) uses the concept of 'kinship' with similar intent. When writing about the Australian Aboriginal context, she describes kinship systems as multifaceted:

> The complexity of Aboriginal kinship systems is well known, and the fact that they include non- humans as well as humans is a further aspect of their complexity. Cross-cutting categories of kin groups generate a system of looped, entangled and enduring moral bonds of care, responsibility, accountability and exchange.
>
> (Rose 2013, 137)

As an example of this interconnectedness in an Irish context, I would like to point to *The Song of Amairgen*, which is widely regarded as the first poem composed in Old Irish, and one that demonstrates a holistic vision of the universe. At the centre of the song is Amairgen Glúingel (a mythical or fictional poet in early Irish literature), who was chief bard of the Sons of Mil. The Milesians (aka the Gaels) invaded Ireland in revenge for the killing of their great-uncle Íth by the three kings of the resident Tuatha Dé Danann. Ultimately there was a battle between the Tuatha Dé Danann and the Milesians. Prior to the battle, the Druids of the Tuatha Dé Danann raised a magical storm to keep the Milesians from reaching the shore. As he set his right foot upon Ireland, Amairgen sang an invocation in response, calling upon the spirit of Ireland. With that, Amairgen was able to part the storm and bring the ship safely to land. In the subsequent battle, there were heavy losses on all sides, but the Milesians eventually won, and the Tuatha Dé Danann were banished to the hollow hills. The text of the song is recorded in the eleventh-century *Lebor Gabála Érenn*, usually known in English as the *Book of Invasions*. John Carey (2023 [1994], 265) translates and annotates *The Song of Amairgen* as follows:

> I am a wind in the sea (for depth).
> I am a sea-wave upon the land (for heaviness).
> I am the sound of the sea (for fearsomeness).
> I am a stag of seven combats (for strength).
> I am a hawk upon a cliff (for agility).
> I am a tear-drop of the sun (for purity).
> I am fair (i.e. there is no plant fairer than I).
> I am a boar for valour (for harshness).
> I am a salmon in a pool (for swiftness).
> I am a lake in a plain (for size).
> I am the excellence of arts (for beauty).

Many scholars have recognised the sense of unity implicit in the poem. Daithí Ó hÓgáin (1991, 23) writes that when 'Amairgen first lands, he recites a great mystical rhetoric in which he exults in being a poet, claiming to be at one with the whole environment. He is wind, sea, bull, hawk, dewdrop, flower, boar, salmon lake, and hill'. Garry MacKenzie (2019) suggests that the song is reminiscent of the concept of 'mesh' as used by Timothy Morton to imply the interconnectedness of all life forms. Morton (2011, 22) defines mesh as follows:

> The mesh is even more deeply interwoven than biocentric ideas such as the web of life imply, because it does away with boundaries between living and nonliving forms. At the same time, symbiosis ensures that boundaries between life-forms are never rigid and thin; rather, they are wide and permeable.

Celtic Scholars Alwyn Rees and Brinley Rees (1961, 99) draw a comparison between Amairgen and the character of Śrī Krishna in the *Bhagavad-Gita*. They describe Śrī Krishna as 'the cosmic juggler or magician and he is all those appearances through which the true essence of existence manifests itself – the cycle of the year, light, wind, earth, weather, the four quarters of space and so on'. Just as Śrī Krishna suggests that he is the divine seed without which nothing would exist, Amairgen 'embodies the primeval unity of all things. As such he has the power to bring a new world into being' (Rees and Rees 1961, 99).

For me, the song is reminiscent of a poem by Native American author and poet Scott Momoday. His poem entitled *The Delight Song of Tsoai-talee* makes the link with his Kiowa Indian name, Tsoai-talee, which means 'rock-tree boy'. When the poet was a baby, an old man who had taken Momaday to a sacred place that his tribe called 'rock-tree' had given him this tribal name. In his poem of delight in being alive, Scott Momoday (1983, 158) begins:

> I am a feather on the bright sky
> I am the blue horse that runs in the plain

In their analysis of the poem, Jasim and Addim (2023, 301–302) suggest that 'this type of poetry is an expression of the deep connection between man, earth, spirituality, and beauty'.

Multispecies

In many traditional folktales in Gaelic, the relationship between humans and non-humans is non-hierarchical. As in most mythologies worldwide, the divide between gods, humans, and animals is blurred. Humans can change into werewolves, hares, insects, and so on. Shapeshifting occurs frequently. The most dramatic example is that of Tuan Mac Carill (from the *Book of Invasions*) who was the sole survivor of the Parthalón settlement of Ireland. Over centuries of survival, he underwent several transformations from stag to boar to hawk, and finally as a salmon, before being eaten and reborn as a human. There are many examples that can be cited in this context.

Morrigan (Mór-Ríogháin), the Goddess of War, could become visible as a woman or animal or bird. In the medieval Irish tale *Táin Bó Cuailgne* (the Cattle Raid of Cooley), Morrigan delivers a warning to *Donn Cuailgne* (the Brown Bull) that Queen Maebh had plans to abduct him. When the Queen's warriors attack Ulster, the province is defended by a single warrior known as Cú Chulainn, who himself undergoes a grotesque transformation. Morrigan attempts to distract Cú Chulainn from battle by turning herself into a beautiful young woman and asking for his love. When Cú Chulainn refuses, she tries an alternative approach by changing into an eel and trapping him. Not satisfied with this attempt, she subsequently becomes a wolf, which serves as a catalyst for a stampede of a herd of cattle towards the lands. Transforming into a heifer, she herself leads this charge of cattle. Her magical shapeshifting from goddess to human to animal does, however, not defeat Cú Chulainn, who succeeds in wounding her at each transformation. Finally, she changes into an old woman – and it is in this human shape that she succeeds in her destructive quest. Milking a cow bearing her three wounds, she offers the champion milk to drink. He accepts it and gives her a blessing which heals Morrigan's wounds. He subsequently regrets this empathy.

Many stories in Celtic mythology are associated with the figure of a deer who is sometimes human. The legendary warrior Fionn Mac Cumhail, who loved to hunt at Slievenamon in Tipperary, marries Sadbh – a woman who has been transformed from a deer into human form when she met Fionn (Nic Craith 2023). When tricked by an evil force, she must revert to the wild as a deer – but not before she gives birth to Oisín, who is subsequently found by his father at Ben Bulben near Sligo. Oisín's name means 'little deer'. In Bateman's words (2009, 144): 'The hero, Fionn, originally a god, changes shape as a dog, deer or man; his nephews are dogs, one of his wives is a deer and his son, Oisean is part-deer'.

On some occasions, the shapeshifting occurred at the behest of the shapeshifter. On others, the transformation serves as a punishment, as in the story of the Children of Lir – a tale associated with Inis Glora and the Erris Peninsula in Mayo. Jealous of the love her husband has for his children, their stepmother turns them into swans for 900 years. They are destined to spend 300 years in Lake Derravaragh, another 300 years on the Sea of Moyle, and a final 300 years in Inis Glora, where they return to their human condition following the intervention of a Christian saint. Stories such as these, which are linked to particular places, give memory to a locality and enhance a sense of belonging to a particular 'patch'. The implications of these stories are the blurring of boundaries between humans and animals. There is no hierarchy. This also gives life to the earth itself.

Place and Narrative

Dúchas encourages the reading of the landscape as an animate being. MacKinnon (2011, 78) writes: 'The idea that the land we live in and belong to is not just a landscape, but a deeply peopled, storied place, is integral to Gaelic and indigenous understandings of the world'. *Dúchas* is best envisaged as a series of concentric circles that extend from the local outwards. This perspective aligns well with Ní Mhathúna's (2021, 251) definition, which explains *Dùthchas* as an 'indigenous cultural concept [...] representing an expanded place-based way of knowing'. In this, the language of the Gaels displays a perspective that has much in common with that of other indigenous peoples.

In Gaelic culture, the land was often regarded as represented by a range of goddesses. The island of Ireland has three distinct manifestations, Éire, Banbha, and Fódhla. The three sisters were representations of the Goddess of Irish sovereignty. Éire is the most familiar sister and is the one whose name was bestowed on Ireland. Examples of patches associated with more local goddesses include the *Dá Chích Anann* (Paps of Anu), which refers to a pair of mountains in Kerry that are said to resemble breasts of Anu (or Danu), who was believed to have been the mother of all gods in Ireland. 'On the top of each of the Paps there is a cairn, possibly a Neolithic burial mound' (Bondarenko 2018, 57).

Similarly, *Dá Chích na Morrigna* in County Meath are known as the Paps of Morrigan, the aforementioned Goddess of War (Ó hÓgáin 1991, 307). Of examples such as these, Bateman and Purser (2020, 284) propose that in the Gaelic imagination:

> the land itself was felt to be animate as represented by a range of goddesses, and that the goddess herself – and hence the land – could manifest herself in any living form, making all life interconnected, plant, animal, human and spirit.

Stories regarding landscape frequently refer to (often small) specific places. The importance of the local is best captured with reference to *Dinnshenchus*, placelore, which references legends and incidents and/or attachments to particular places. Chiblow and Meighan (2022, 208) say in relation to Scottish-Gaelic that:

> an intimate knowledge of the land is embedded in Indigenous languages. For example, in Scotland, there are places in Gàidhlig that speak to the landscape and what can typically be found or encountered there: *Alltan Èisg* ('small burn/stream of fish'), *Fèith Gaineimh Mhòr*, ('the big sandy bog-stream'), or *Loch nam Breac Mòra*.
>
> ('loch of the big trout')

Connection with the land engenders a strong sense of place among Irish people speaking the indigenous language. This sense of place can be compared with that described by Keith Basso in his *Wisdom Sits in Places* (1996) where he explores the role of placenames in the language and culture of the Western Apache. Basso argues that Western Apache way of life can only be understood by examining the people's sense of place. He lays out the importance of place-making as a cultural activity, one that involves both imagination and memory, and is inherently focused on the local. It is a form of place-making that is integrally linked with the indigenous language.

Alienation and Landscape

Placelore and placenames in the indigenous language are manifestations of the relationship between people, species, and the landscape (Nic Craith 2020). Unfortunately, the Irish language is no longer mainstream in Ireland today. While one can point to different factors in the decline of the language, chief among them is the internal colonisation of Ireland by forces that can be dated back to Anglo-Norman invaders in 1169. This marked the beginning of more than 800 years of foreign domination, which lasted until 1922 when Ireland was partitioned, and the South of Ireland became independent from Britain (Beckett 1966; also, Moody and Martin 1967). Porter (2010) suggests that the experience of English colonists in Ireland informed the Spanish subjugation of the Americas. This comparison is also endorsed by Canny (1988), who hypothesises that Ireland was an experiment for the colonisation of North America.

The colonisation process impacted both landscape and language. Tsing et al. (2019) point out that European colonialism was reliant on landscape structures such as deep-water harbours and resource plantations. Capitalism on a global scale spread fossil fuel-burning factories and container shipping. The Great Acceleration intensified in the decades after the Second World War. In Ireland, part of the process of colonisation was the 'civilization' of Irish-speakers, who were strongly encouraged to give up their native language and adopt English instead (Nic Craith 1993).

Anglicisation and re-naming of places in the language of the coloniser occurred in many countries worldwide (Nic Craith 2012). Spivak cites the example of a nineteenth-century British soldier travelling across India. On his journey, he surveys the land, which is now under British imperial rule. The process involves more than the translation of words and names. The British soldier is consolidating British rule by obliging the native inhabitants to disconnect from the local placename. It is now the British placename that describes their home ground. 'He [the soldier] is worlding their own world, which is far from mere uninscribed earth, anew, by obliging them to domesticate the alien as Master' (Spivak 1985, 253).

The colonial era in Ireland hastened the decline of original Irish placelore and placenames. The process of translation severed people from the meaning of the land. The process of Anglicisation ruptured the reservoir of local folklore and the transmission of traditional ecological knowledge. The connection between land and languages is highlighted by Michael Cronin (2019, 9), who argues that 'language situates people in their environment in terms of both description and narration. It tells you where you are and what's around you. The removal of Irish from public life has alienated people from their own surroundings'.

This sense of alienation from the landscape is powerfully captured in Brian Friel's (1984) play *Translations*. This celebrated theatre production, first performed in Derry, explores the ordnance survey of Ireland, carried out by the British Army Engineer Corps in the 1830s. The action takes place in August 1833 in a fictional village called *Baile Beag* (little town) located in County Donegal in the North-West of Ireland. Owen, the son of the hedge-school[2] master, helps the British army engineer Yolland with the mapping and re-naming of places in English. The play devotes considerable attention to the historical and cultural significance of the 'translation' of placenames. In fact, the placenames were not really translated. Instead, they were changed to nonsense words in English that sounded like the original Irish. While technically a 'simple' administrative process, the consequences were enormous, as it severed the people from the stories and folklore associated with the original Irish placenames. As an example of this we can look at the placename *Tobair Bhriain* (Brian's Well). In the play it is changed to the meaningless Tobair Vree. Owen explains the original story about Brian and the well (Friel 1984, 420):

> And we call that crossroads Tobair Vree. And why do we call it Tobair Vree? I'll tell you why. Tobair means a well. But what does Vree mean? It's a corruption of Brian – (*Gaelic pronunciation*) Brian – an erosion of Tobair Bhriain. Because a hundred-and-fifty years ago there used to be a well there, not at the crossroads, mind you – that would be too simple but in a field close to the crossroads. And an old man called Brian, whose face was disfigured by an enormous growth, got it into his head that the well was blessed; and every day for seven months he went there and bathed his face in it. But the growth didn't go away; and one morning Brian was found drowned in the well. And ever since that crossroads is known as Tobair Vree – even though the well has since dried up.

In Ireland, the process of Anglicisation occurred at the same time as the landscape was losing the species that inspired some of its Irish placenames. Bateson (2021a, 116) suggests that many indigenous placenames are a 'ghost map of former species. In these cases, the wild boar still haunts *Ceann Toirc*, the eagle *Gleann an Íolair* and the wolf *Faol Droim*'. Examples from Northern Ireland include Legoneil (translated from *Lag an Aoil,* meaning 'Hollow of the Lime') on Wolf Hill. The name refers to the site near Belfast where the last wolf in Ireland was killed (McCoy 2019). One could easily fall into despair about the traditional ecological knowledge that has been lost.

Radical Hope and Cultural Darning

This sense of despair has been described by Deborah Bird Rose as a 'double death' (2004, 176). When White Australian settlers arrived on the continent, they brought with them a very specific, linear concept of time that contained a key characteristic of modernity – the separation of nature from culture. This concept of 'double death' aptly describes the separation of nature from culture in an Irish context. It was to counter this disillusionment that Tsing et al. (2019) designed the concept of 'Patchy Anthropocene'. Instead of dealing with helplessness on a global scale, activities could zone in on specific places with an attitude of optimism and ambition.

With this approach, Tsing and others were responding to Lear's concept of 'radical hope'. In his book with that title, Lear (2008) explores the situation of a tribe confronted with annihilation. Lear's case study was the life-narrative of a Crow chief named Plenty Coups. When describing the disappearance of the buffalo herds on the plain, Plenty Coups remarked to Frank Linerman: 'But when the buffalo went away the hearts of my people fell to the ground, and they could not lift them up again. After this nothing happened' (Lear 2008, 2). Identification with warrior status was hugely significant for young Crow men. The loss of the buffalo resulted in a huge sense of disorientation in a context of poverty, disease, and dislocation. In response to this context, Lear posed the question of whether it is possible for indigenous groups to re-invent their 'webs of significance' (Geertz 1973, 5) and cultivate a sense of optimism – or was it inevitable that the legacy of colonisation would generate 'death by 1,000 cuts'? Lear argued that while the latter option might seem inevitable, the former (more hopeful) option was also feasible. One could acknowledge catastrophe but also imagine a radically transformed future. This would occur – not at planetary or global level – but in local patches.

Drawing on the experiences of Indigenous peoples who have previously faced deforestation and annihilation, one can imagine local solutions that are permeated with a radical courage that is 'directed towards a future goodness that transcends the current ability to understand what it is' (Lear 2008, 103). This solution will not be achieved by a single mode of knowing favoured by Western society. Instead, it will involve different modes of knowledge (Tsing et al. 2019). Although alternative ways of knowing such as *dúchas* have been diluted, erased by the colonial project, such indigenous modes of knowledge are crucial if we are to invest localities with relevant thinking (Tsing et al. 2019).

In keeping with the concept of reclaiming a traditional place-based perspective, it is appropriate for Gaels to revitalise the concept of *dúchas* and 'the *re-indigenisation* of Irish language politics in the age of the Anthropocene' (Cronin 2019, 11). The aim of this regeneration would be to reclaim the 'ontological dynamic of embodied experience and terms and emplacement ("on the ground"), and complex entanglement ("in the mind") with relationships of belonging and dwelling, heritage and inheritance, a human ecology with "place" (including, where relevant, land)' (Meighan 2022). That is understood as 'political, social and cultural imagination in action' (Oliver 2021, n.p.). Such rekindling of identification with the land would generate 'tangible action and conduct' and a sense of responsibility for a particular 'patch of land', but the focus on locality is neither exclusive nor inward looking. This local perspective is sometimes mis-interpreted as a form of populism, nationalism, or xenophobia. Instead, in the words of Mairi McFadyen (2019, n.p.):

> [A]dvocating for local culture is not about reifying places and forms of non-capitalism as untouched or outside of history as part of some sort of romantic hankering for paradise lost, it is to stand up against the destructive and homogenising forces of capitalist modernity.

This is in the spirit of Scottish ecologist Patrick Geddes, who advised people to 'think global and act local' (MacDonald 2020). Re-engaging with the local is a radical approach that favours multiplicity, fluidity, and contradictions rather than a monolithic, static approach. It is cross-generational. It reaches back to previous cultural memories, but also thinks forward about future generations, in a spirit of democracy, 'working towards a shared vision of a thriving, equitable and convivial society' (McFadyen and Sandilands 2021, 166). Such a forward-looking agenda will involve discussion as to how the Gaelic concept *dúchas* can be mobilised to envision an alternative nature–culture future.

This approach might involve what McFadyen and Sandilands (2021) call 'cultural darning and mending', the elements of which are visualised in Figure 7.2.

Figure 7.2 Cultural darning. Figure by Mairéad Nic Craith.

They describe this as a 'playful approach to cultural activism', or a future-orientated creative ethnology that engages with maps, placenames, and local knowledge. In linking to the past, we 're-member' (McIntosh 1988) a *dúchas*-grounded relationship with the land – and raise consciousness about what was lost during the process of Anglicisation and colonisation.

As an example of this, Sandilands points to the engagement of young Scottish-Gaelic children with Disney characters but despairs that they lack knowledge about pan-Celtic super-heroes, such as Fionn MacCumhail. This is a scenario that resonates strongly with me. In the early stages of my academic career, I was studying Old Irish as a night student in University College Cork while also teaching young children during the day at a local primary school. At university we were engaging with a text concerning Cú Chulainn (the Hound of Ulster). As a way of familiarising myself with the text, I started to read it (in translation) to the class of six-year-olds that I was teaching. Their fascination with the Celtic character was immediate and powerful. They loved the 'warp-spasm' that seized the character of Cú Chulainn and turned him into a hideous monster. It affected his shanks and joints. Under his skin, his whole body became twisted. One of his eyes fell out onto his cheek while the other sunk deeply into his skull. His mouth warped and his cheeks became disfigured ... At the time of my telling the story, the series *The Incredible Hulk* was running on Irish television. The children readily compared Cú Chulainn with the Incredible Hulk, but found the Celtic character much more interesting as his monstrosity was constructed *by* them in their own imagination, rather than *for* them on the television. The traditional folktale was as mesmerising as any Disney creation.

A re-engagement with folktales is not simply for the purpose of hearing a story. Linking tales with particular places builds a sense of empathy and connection. It also has the potential to highlight ecological knowledge about particular places that can serve as pointers for ecological survival (Cronin 2019, 12). Bateson (2021b) suggests that, in Ireland, our indigenous language is the most appropriate channel to re-invigorate our *dúchas*. Irish is in an unusual position in terms of its linguistic status. It is recognised as the official and first national language of Ireland. It also has status as working language of the European Union. As such, it has associations with capitalism and toxicity. Bateson (2021b, 52) notes that Irish 'is a native language of a wealthy Western European country'. Despite this status, it is not widely spoken throughout the country and effectively operates as a minority language. It is a language that has been repressed during a process of colonisation that resulted in its retreat to rural locations where it maintained its historical empathy with the landscape. The folklore collected in early twentieth-century Ireland shows 'how isolated communities interacted with the non-human before the widespread arrival of electricity or the car' (Bateson 2021b, 52). For this reason, Bateson (2021b) argues that Irish was less implicated with modernism than majority languages such as English.

In his *We Have Never Been Modern*, Latour (1993) characterises modernity as having bolstered a divide between nature and culture. Given that repressed languages such as Irish were considered 'anti-establishment, anti-colonial', perhaps they have the potential to subvert the modernist human vs animal paradigm and restore a multispecies perspective. 'Through their exclusion from industrial modernity and new capitalism, minority languages and the culture produced in them might offer another way to examine human/non-human relations' (Bateson 2021b, 52). This is a flat rather than a hierarchical perspective – one that has been termed by philosopher Graham Harman (2018, 9) as Object Orientated Ontology ('OOO' or triple 'O'). Basically, the world is viewed from a flat perspective and 'all objects must be given equal attention whether they be human, non-human, natural, cultural, real or fictional' (Harman 2018, 9). This is a reversal of the Enlightenment project that gave the human authority over nature, which had a very negative effect. A revised perspective would regard humans as no more or less important than any other element of nature.

Conclusion

This chapter has explored the concept of *dúchas* with a view to exploring its potential to nurture a renewed nature–culture relationship in local places in Ireland. Given the type of relationship implied in *dúchas*, the revival of this indigenous concept in local places has the potential to facilitate and generate a greater appreciation of the role of the human as part of (rather than in charge of) nature. 'Indigenous languages transmit unique ways of understanding and relating to the world. Safeguarding Indigenous languages and knowledges is vital for addressing the current climate and humanitarian crises' (Chiblow and Meighan 2022, 209). In pre-Enlightenment Ireland, there was strong empathy with local places and the multiplicity of species that lived there. This connection was reinforced through Gaelic narratives and placelore. With Anglicisation, that connection was disrupted but not completely lost. It is not yet too late to revisit and reclaim some of the original meaning of *dúchas*. A recovery of the future-facing meaning of *dúchas* is best facilitated by land-based learning. This is not to advocate a wholescale reversion to the ontology that dominated pre-colonial Ireland – but it is to propose a cultural transformation that would address the issue of climate change in Ireland in a manner that engages local, place-based resources.

Notes

1 I am using 'indigenous' here in the sense implied in the following definition: 'Derived from old Latin, it means "born or produced from within," with primary definitions suggesting nativeness; originating or growing in a country; not exotic' (Clifford 2013, 13–14).

2 Hedge schools were unofficial education initiatives during parts of the eighteenth and nineteenth centuries, effectively subverting colonial restrictions on schooling.

Bibliography

Adams, Carol and Lori Gruen, ed. 2013. *Ecofeminism: Feminist Intersections with Other Animals and the Earth*. London: Bloomsbury.

Basso, Keith. 1996. *Wisdom sits in Places: Landscape and Language among the Western Apache*. Albuquerque: University of New Mexico Press.

Bateman, Meg. 2009. "The Landscape of the Gaelic Imagination." *International Journal of Heritage* 15 (2–3): 142–152.

Bateman, Meg and John Purser. 2020. *Window to the West: Culture and Environment in the Scottish Gàidhealtachd*. Sleat: Clò Ostaig.

Bateson, Callum. 2021a. "Folklore and the Climate Crisis: Reading Beara as an Anthropocene Patch with Máiréad Ní Mhíonacháin." *Networking Knowledge* 14 (2): 110–123.

Bateson, Callum. 2021b. "Resurrecting Alltar: Looking Past the Anthropocene with Liam Ó Flathearta's Dúil." *Imaginaires* 23: 48–57.

Beckett, James. 1966. *The Making of Modern Ireland 1603–1923*. London and Boston, MA: Faber and Faber.

Bondarenko, Grigory. 2018. "Ireland as Mesocosm." In *Celtic Myth in the 21st Century: the Gods and their Stories in a Global Perspective*, edited by Emily Lyle, 53–71. Cardiff: University of Wales Press.

Canny, Nicholas. 1988. *Kingdom and Colony: Ireland in the Atlantic World, 1560–1800*. Baltimore and London, MD: John Hopkins University Press.

Carey, John. (1994) 2023. "The Book of Invasions." In *The Celtic Heroic Age: Literary Sources of Ancient Celtic Europe and Early Ireland and Wales*, edited by John Koch and John Carey, 226–271. Aberystwyth: Celtic Studies Publications.

Chiblow, Susan and Paul Meighan. 2022. "Language is Land, Land is Language: The Importance of Indigenous Languages." *Human Geography* 15 (2): 206–220.

Clifford, James. 2013. *Returns: Becoming Indigenous in the Twenty-First Century*. Cambridge, MA: Harvard University Press.

Cronin, Michael. 2019. *Irish and Ecology/An Ghaeilge agus an Eiceolaíocht*. Baile Átha Cliath: Foilseacháin Ábhair Spioradálta.

Crutzen, Paul and Eugene Stoermer. 2000. "The 'Anthropocene'." *Global Change Newsletter* 41: 17–18.

D'Eaubonne, Françoise. 2022. *Feminism or Death: How the Women's Movement Can Save the Planet*. London: Verso.

Friel, Brian. 1984. *Selected Plays*. London and Boston, MA: Faber and Faber.

Gaard, Greta. 1993. *Ecofeminism: Ethics and Action*. Philadelphia, PA: Temple University Press.

Garry MacKenzie. 2019. "Ben Dorain: An Ecopoetic Translation." *Humanities* 8(2), http://dx.doi.org/10.3390/h8020113

Geertz, Clifford. 1973. *The Interpretation of Cultures*. New York: Basic.

Haraway, Donna. 2015. "Anthropocene, Capitalocene, Chthulucene. Donna Haraway in Conversation with Martha Kenney." In *Art in the Anthropocene: Encounters among Aesthetics, Politics, Environments and Epistemologies*, edited by Heather Davis and Etienne Turpin, 255–270. London: Open Humanities.

Harman, Graham. 2018. *Object-Oriented Ontology: A New Theory of Everything*. London: Pelican.

Hechter, Michael. 1975. *Internal Colonialism: The Celtic Fringe in British National Development 1536–1966*. London: Routledge and Kegan Paul.

Jasim, Mustafa and Addim Soukaina. 2023. "Native American Spirituality and Natural Landscape in N. Scott Momaday's Poetry." *Language Literacy: Journal of Linguistics, Literature, and Language Teaching* 7 (2): 292–304.

Kimmerer, Robin. 2020. *Braiding Sweetgrass*. London: Penguin.

Latour, Bruno. 1993. *We Have Never Been Modern*. Cambridge, MA: Harvard University Press.

Lear, Jonathan. 2008. *Radical Hope: Ethics in the Face of Cultural Devastation*. Cambridge, MA: Harvard University Press.

MacDonald, Murdo. 2020. *Patrick Geddes' Intellectual Origins*. Edinburgh: Edinburgh University Press.

MacKinnon, Iain. 2011. 'Eachdraidh nar Cuimhne' – History in our Memories: an Analysis of the Idea that the Highlands and Islands of Scotland can be understood as a Site of Colonisation. [Unpublished Doctoral Dissertation]. Ulster University. Accessed 4 June 2024. https://ethos.bl.uk/OrderDetails.do?uin=uk.bl.ethos.573684

MacKinnon, Iain. 2019. "Education and the Colonisation of the Gàidhlig Mind." *Bella Caledonia*. Accessed 10 January 2024. https://bellacaledonia.org.uk/2019/12/03/education-and-the-colonisation-of-the-gaidhlig-mind/

MacKinnon, Iain and Ruth Brennan. 2012. *Dùthchas Na Mara/Belonging to the Sea: Exploring the Cultural Roots of Maritime Conflict on Gaelic Speaking Islands in Scotland and Ireland*. Alba: SAMS/Scottish Crofting Federation.

McCoy, Gordon. 2019. *Gaelic Maps of Belfast*. Belfast: Her Majesty's Stationery Office.

McFadyen, Mairi. 2019. "Songs and Singing: A Wellspring for the Future." *Bella Caledonia for The National*. Accessed 20 September 2024. https://www.mairimcfadyen.scot/blog/2019/12/27/songs-amp-singing-a-wellspring-for-the-future-bella-caledonia-column-december-2019

McFadyen, Mairi and Raghnaid Sandilands. 2021. "Cultural Darning and Mending: Creative Responses to Ceist an Fhearainn/The Land Question in the Gàidhealtachd." *Scottish Affairs* 30 (2): 157–177.

McIntosh, Alistair. 1988. The Gal-Gael Peoples of Scotland: on Tradition Re-bearing, Recovery of Place and Making Identity Anew. Accessed 20 September 2024. https://www.alastairmcintosh.com/articles/1998_galgael_htm.htm

McNeill, John and Peter Engelke. 2016. *The Great Acceleration: An Environmental History of the Anthropocene since 1945*. Cambridge, MA: Harvard University Press.

Meighan, Paul. 2022. "Dùthchas, a Scottish Gaelic Methodology to Guide Self-Decolonization and Conceptualize a Kincentric and Relational Approach to Community-Led Research." *International Journal of Qualitative Methods* 21: 1–14.

Moody, Theodore, Francis Martin and Dermot Keogh, ed. 1967. *The Course of Irish History*. Dublin: Raidió Teilifís Éireann.

Morton, Timothy. 2011. "The Mesh." In *Environmental Criticism for the Twenty-First Century*, edited by Stephanie LeMenager, Teresa Shewry and Ken Hiltner, 19–30. Routledge: New York.

Murray, John. (2014) 2019. *Reading the Gaelic Landscape: Leughadh Aghaidh Na Tìre*. Dunbeath: Whittles.

Newton, Michael. 2019. *Warriors of the Word: The World of the Scottish Highlanders*. Edinburgh: Birlinn.

Nic Craith, Mairéad. 1993. *Malartú Teanga: An Ghaeilge i gCorcaigh sa Naoú hAois Déag*. Bremen: European Society for Irish Studies.

Nic Craith, Mairéad. 2012. *Narratives of Place, Belonging and Language: An Intercultural Perspective*. London: Palgrave Macmillan.

Nic Craith, Mairéad. 2020. *The Vanishing World of the Islandman: Narrative and Nostalgia*. Bern: Springer.

Nic Craith, Mairéad. 2023. "Gaelic Folklore for a Multi Species Future." *Bella Caledonia*. Accessed 30 October 2023. https://bellacaledonia.org.uk/contributor/mairead-nic-craith/

Nic Craith, Mairéad. 2025. "The Multi-Species Paradigm in Traditional (Gaelic) Folklore." In *Narrating the Multispecies World Stories in Times of Crises, Loss, and Hope*, edited by Michaela Fenske, 233–246. Bielefeld: Transcript. https://creativecommons.org/licenses/by-sa/4.0/

Ní Mhathúna, Deirdre. 2021. "Traditional Ecological Knowledge and the Relevance of Dùthchas in Gàidhealtachd Environmental Futures." *Scottish Affairs* 30 (2): 251–261.

Ó hÓgáin, Daithí. 1991. *Myth, Legend and Romance: An Encyclopaedia of the Irish Folk Tradition*. New York: Princeton Hall.

Oliver, James. 2021. "Acknowledging Relations: Dùthchas, Seanchas and Ethical Emplacement." In *The Commonplace Book of ATLAS*, edited by Emma Nicolson and Gayle Meikle, 30–33. Portree: ATLAS Arts.

O'Loughlin, Michael. 2008. "Radical Hope, or Death by a Thousand Cuts? The Future for Indigenous Australians." *ARENA* 29/30: 75–201.

Porter, Joy. 2010. "Primitive Discourse: Aspects of Contemporary North American Indian Representations of the Irish and of Contemporary Irish Representations of North American Indians." *American Studies* 49 (3/4): 63–85.

Rees Alwyn and Brinley Rees. 1961. *Celtic Heritage: Ancient Tradition in Ireland and Wales*. London: Thames and Hudson.

Robinson, Tim. 2003. "The Irish Ecosphere." *New Hibernia Review/Iris Éireannach Nua* 7 (3): 9–22.

Robinson, Tim. (1996) 2007. *Setting Foot on the Shores of Connemara and Other Writings*. Dublin: Lilliput.

Rose, Deborah. 2004. *Reports from a Wild Country: Ethics for Decolonisation*. Sydney: University of New South Wales.

Rose, Deborah. 2013. "Death and Grief in a World of Kin". In *The Handbook of Contemporary Animism*, edited by Graham Harvey, 137–147. Stocksfield: Acumen.

Scott Momoday, Navarre. 1983. "The Delight Song of Tsoai-talee". In *Songs from the Earth on Turtle's Back: Contemporary American Poetry*, edited by Joseph Bruchac, 158. New York: Greenfield Review.

Smith, Iain Crichton. 1982. *Towards the Human*. Loanhead: Macdonald.

Spivak, Gayatri. 1985. "The Rani of Sirmur: An Essay in Reading the Archives." *History and Theory* 24: 247–272.

Tsing, Anna, Nils Bubandt, Elaine Gan and Heather Swanson, ed. 2017. *Arts of Living on a Damaged Planet: Ghosts and Monsters of the Anthropocene*. 3rd ed. Minneapolis: University of Minnesota Press.

Tsing, Anna, Andrew Mathews and Nils Bubandt. 2019. "Patchy Anthropocene: Landscape Structure, Multispecies History, and the Retooling of Anthropology." *Current Anthropology* 60 (20): 186–197.

Wierzbicka, Anna. 1997. *Understanding Cultures through their Key Words: English, Russian, Polish, German, and Japanese*. Oxford: Oxford University Press.

8 Rural Entrepreneurship, Locality, and Future-Sustainable Environment

Maarit Grahn

Tourism industries and tourism development have become a key means of preserving the vitality of rural and remote areas as the economic structure changes (Hakkarainen 2017). In the transformation of the countryside, cultural heritage is used more and more as a resource for tourism businesses. In recent years, the development of tourism has especially emphasised intangible living heritage, such as local customs, festivities, food, and nature-related knowledge (Malkamäki and Tulila 2012).

One prevailing trend in the tourism is locality, which can be considered as so-called new authenticity. Tourists want to do the same things that locals do (Kaipainen 2015). In the development of tourism industries, more and more attention is also paid to sustainable development. The goal of sustainable development is to generate economic benefits for local communities and promote the preservation of cultural heritage (Lordkipanidze et al. 2005).

The role of private actors in regional development and cultural heritage work is constantly growing. One reason for this development is the continuous cost-saving measures of the public sector (Brulin 2007; Ilmonen 2016). The aim of this chapter is to emphasise the influence of rural entrepreneurship in maintaining and transmitting traditions and values associated with locality, human–nature relationships, and future-sustainable environment. The focus is especially on nature-based entrepreneurship. I concentrate on the entrepreneurs' values and attitudes associated with nature and sustainable development, and on the driving factors for starting as an entrepreneur. I also examine how the features of nature (e.g., rivers, rapids, springs, maritime environments, forests, rocks, fields, or animals), and the local stories associated with these features, are utilised, and what kind of practices entrepreneurs have adopted in order to advance sustainable development.

The rural entrepreneur is often a tourism entrepreneur. I will approach rural entrepreneurship from the point of view of some entrepreneurs operating in Merikarvia. Merikarvia is a municipality of about 3,100 inhabitants located in Satakunta, on the coast of the Bothnian Sea. The main local sources of livelihood are small industries, service trades, and tourism, which is constantly growing. Merikarvia also has more than 1,600 leisure apartments,

DOI: 10.4324/9781003727972-8

which is why summer residents are, along with tourists, an important target group as users of local services (see, e.g., Sivula et al. 2017).

The local cultural heritage of Merikarvia ranges from rural, coastal, and natural heritage to the cultural heritage of business or industry. For centuries, the sea has enabled the local population to make a living through fishing, seal hunting, trade, port operations, and building boats and ships. In the nineteenth century, the Merikarvianjoki river was an important passageway, a log floating route, and a source of power for mills and sawmills. Nowadays, the local cultural heritage is strongly connected to the sea, the river, and fishing, and this is also reflected in the way in which the cultural heritage is utilised in the tourism industry. The Bothnian Sea National Park and the Oura archipelago are scenically, biologically, and also historically significant areas. The Merikarvianjoki river is one of the most popular fishing destinations in Southern Finland. The length of the fishing permit area is about 24 km, and the river route has numerous rapids along which mills or sawmills used to operate. Some of the rapids are even named after them. Along with the sea and the river, the stone fences and traditional standing stones (*tonttukivet*) are also typical elements of local cultural heritage (see, e.g., Sivula et al. 2017; Merikarvianjoki 2022; Metsähallitus 2024).

This chapter is related to my research that examines cultural heritage entrepreneurship in promoting vitality, sustainable development, and local identity. The chapter is connected to discussions about cultural heritage and rural entrepreneurship and cultural heritage processes (Sivula 2009; Brulin 2007; Ilmonen 2016; Lordkipanidze et al. 2005; Kaipainen 2015; Korsgaard et al. 2015). In rural areas, nature creates local cultural heritage as well as content and boundaries for tourism activities (Hakkarainen 2017). The relationship between human and nature, and knowledge related to nature, is part of living, intangible cultural heritage (Siivonen 2019; Hirvonen 2017; Sievänen 2017). Therefore, nature-based, tourism-focused entrepreneurship can also be viewed from the perspective of cultural heritage entrepreneurship: nature-based entrepreneurship is maintenance of a relationship with nature, maintenance and transmission of nature-related traditions and knowledge.

I start this chapter by presenting the research data and the data analysis method. After this, I continue with the presentation of the theoretical framework. I introduce previous research on rural entrepreneurship and cultural heritage entrepreneurship and bring out some terminology related to the topic. The next three sections form the empirical part of this chapter. First, I examine the values and motives of rural entrepreneurs in the light of the research material. After that, I examine what kind of cultural heritage entrepreneurs refine into tourism products and services, and how they refine it. The next section introduces sustainable development, and the role of entrepreneurs as mediators of cultural heritage and builders of local identity. Finally, I present the most important conclusions.

Data and Method

The research data consists of thematic interviews with the entrepreneurs, and observations concerning the activities of the entrepreneurs, which I carried out during 2020–2022. For my research, I interviewed ten local entrepreneurs.[1] For this chapter I have used only part of the collected data because the text focuses mainly on nature-based entrepreneurship.

One of the interviewees is a nature tourism entrepreneur whose services include nature and canoeing trips and cabin accommodation services (Interview 1). The next entrepreneur is engaged in breeding Finnish horses, and she also organises riding lessons, riding camps, and cross-country hikes for riders (Interview 2). Two of the informants are former fishermen who handle boat transports during summertime to the Oura archipelago in the Bothnian Sea National Park, where the old pilot station of Oura-islet is situated (Interviews 3 and 4). In my study, I call them skippers. There is also a family business which has been engaged in farm tourism for 30 years. The company has undergone a generational change in the autumn of 2020 (Interview 5).

The main topics of the interviews were the values of entrepreneurs and the importance of cultural heritage for entrepreneurs, the factors that influenced starting as an entrepreneur, the ways of utilising cultural heritage in business operations, sustainable development, and local and regional co-operation networks.

Through participant observation, I collected additional material, especially on how cultural heritage is utilised and transmitted in practice. This meant participating in events organised by the entrepreneurs. In spring 2020, the COVID-19 epidemic had a slight effect on the acquisition of material, because some of the events were either cancelled or postponed due to the epidemic. In the summer, however, I managed to observe the activities of the entrepreneurs, and I continued the observations during the autumn. I took plenty of photos and took notes both during and after the visits. In addition, I talked with the entrepreneurs when the visiting situations raised additional questions. The entrepreneurs knew in advance of my arrival and, depending on the situation, I also told the other guests about my presence.

I analysed the data using qualitative research methods. The aim of qualitative research is to increase understanding, enable different interpretations, and give things meaning (Gadamer 2004). The process typical of qualitative research has also been characteristic of this research. I analysed the material during the research process at several stages, at the same time as I collected and analysed the data. Based on the research material, my aim was to form a picture of the characteristics of rural entrepreneurship, the values and motives of entrepreneurs, ways of utilising cultural heritage in the development of countryside and rural tourism, and in building local identity. The key goal of the study was to understand and highlight the point of view of entrepreneurs.

Rural Entrepreneurs – Nature-based Entrepreneurs

Entrepreneurship is a key resource for economic development: it creates growth and acts as a driver of innovation and change. According to economist Joseph Schumpeter's classic definition, an entrepreneur is a risk taker, innovator, and leader (Lordkipanidze et al. 2005). There is always a certain profit logic associated with entrepreneurship: the goal of the activity is economic profitability, and also one of the goals may be the promotion of local vitality.

In rural studies (see, e.g., Korsgaard et al. 2015), the ideal type of entrepreneurship is rural entrepreneurship. Rural entrepreneurship contributes to local development and includes the possibility of optimal use of the resources of the rural area. All the entrepreneurs I interviewed can be considered rural entrepreneurs. Characteristic of rural entrepreneurs is that they utilise the natural, cultural, or historical elements of the area, and often these elements are mixed in entrepreneurs' activities. Rural entrepreneurs are attached to local communities, and they often avoid competition with other local companies. Economic incentives and profit may be of less interest than personal social and cultural aspirations. Location and local resources are competitive advantages for rural entrepreneurs. The operating environment and the products cannot be replicated directly elsewhere. Entrepreneurship cannot be moved from its roots to another place without losing some or all of its values (Kaipainen 2015; Korsgaard et al. 2015; Ilmonen 2016). For instance, the former fishermen's knowledge of the sea, the archipelago, and Oura archipelago's history is so strong that their activities cannot be moved elsewhere without losing some valuable cultural heritage elements.

As Gheorghe Zaman (2015) has argued, entrepreneurs who utilise and refine cultural heritage can be called cultural heritage entrepreneurs. These entrepreneurs produce goods and services, organise and run cultural heritage in a commercial or non-profit manner, depending on the characteristics of cultural heritage. In turn, Göran Brulin (2007) has introduced two interesting terms that also can be applied to my interviewees: Identity Entrepreneurship – Identity Entrepreneur. These terms refer to entrepreneurship, where the entrepreneur refines the local cultural heritage with one's entrepreneurial contribution. An entrepreneur finds resources in cultural heritage and turns them into a business. At the same time, the entrepreneur creates and maintains regional identity and attractiveness with the help of cultural heritage. Rural and nature-based entrepreneurship can also be viewed from this perspective, because nature is part of the cultural environment. Nature is an important identity builder, a place for recreation, and a resource (Siivonen 2019; Hirvonen 2017; Sievänen 2017).

The entrepreneurs I have interviewed represent small businesses. According to previous studies (Lordkipanidze et al. 2005; Zaman 2015; Yachin 2019), small companies have a significant role in rural tourism and in nature-based entrepreneurship. One of the success factors of small companies is their flexibility. Doing business in the countryside is different from in the cities, where the entrepreneur can rely on larger customer flows. Therefore, in rural areas, entrepreneurial activities based on cultural heritage often expand, and other activities may be added to it, for example, property maintenance or holiday cottage rental (see, e.g., Molin 2007). This is often a prerequisite for the continuation of business operations in order to smooth out seasonal fluctuations in tourism. The flexibility of small companies can also be seen in the fact that companies are able to react fairly quickly to new demand or changed circumstances (Lordkipanidze et al. 2005).

Values and Motives of Rural Entrepreneurs

> I enjoy being in nature. When I see those old paths, I can imagine how the people in the past have worn birch bark shoes and walked along the trails. My uncle was a good teacher. We used to walk a lot in the woods and watch many interesting things, like trees and rocks. Sharing the experiences is very important to me.
>
> (Interview 1)

According to previous studies (Lordkipanidze et al. 2005; Lithander 2002), entrepreneurs need forward-directing factors in order to operate. Such factors include motives and the backgrounds or conditions that guide an individual's actions and decisions. Motivations for entrepreneurship can be an individual's need for achievement, desire for independence, job satisfaction from work, role of an entrepreneur, or risk-taking. Often, the reasons to start as an entrepreneur can be something other than economic merits.

According to the entrepreneurs interviewed, interest in one's own field and the need for self-fulfilment are the most important motivations for entrepreneurship. In family business, the choices have also been influenced by the culture and traditions within the family: the new generations have grown up following their parents in both entrepreneurship and in ways of nurturing cultural heritage (Interviews 1–5).

A certain spirit of creativity is mentioned as one dimension of cultural heritage entrepreneurship, where cultural values are central alongside economic values (Zaman 2015). Nurturing cultural values or cultural heritage can act as a central motive for entrepreneurship. The entrepreneur sees cultural heritage as a factor that promotes entrepreneurship, not as an obstacle to it (Brulin 2007). In my study, I examined how cultural values are reflected in the motives and actions of rural entrepreneurs. Cultural values emerged strongly in the answers of the interviewees. Locality, history, nature, and cultural heritage are important to all entrepreneurs, and the will to preserve these elements is reflected in the values and motives of entrepreneurs (Interviews 1–5).

The research also revealed that economic incentives are not necessarily as important for rural entrepreneurs as social and cultural aspirations (see, e.g.,

Korsgaard et al. 2015). On the other hand, some entrepreneurs thought that more attention should be paid to the economic side. According to previous studies, one of the challenges related to the economic use of cultural heritage may be combining cultural values and entrepreneurial spirit (Bowitz and Ibenholt 2009). However, economic and cultural values must not be mutually exclusive. Business operations must be viable, because this is the only way to promote the region's vitality (see, e.g., Grahn 2021).

The entrepreneurs involved in my research are very different in terms of their education and work history. Changes in life situations and self-employment through entrepreneurship also reflect in career choices. Some of the entrepreneurs have previously worked in a completely different profession, and some have retrained before becoming entrepreneurs (Grahn 2022). On the other hand, my research also confirms how the previous work history and experience can be resources that can be utilised in changed circumstances. A good example of this is the former fishermen handling sea transportations to the Oura archipelago:

> I stopped working as a fisherman because the seals hindered too much fishing. The boat was ready, and there was nothing else that made sense. [...] I fished for 27 years.
>
> (Interview 3)

> I also worked as a fisherman, but in 1990 I got a job at the port of Pori and I have been working there ever since. But the sea has always fascinated me, and when I worked as a fisherman in the early 1980s I transported people to the sea even then. I also had the boat ready and it was inspected and could be used for transportation. So, I have continued transporting people as a hobby to this day.
>
> (Interview 4)

An entrepreneur refining cultural heritage can be a native, but this is not necessary (Brulin 2007). This also comes up in the case of Merikarvia. Some of the interviewed entrepreneurs are natives, but some have moved there from elsewhere. Those entrepreneurs who have lived there for long time probably have a stronger relationship with the place. However, business activity itself, refining cultural heritage and harnessing new ideas for the use of tourism does not depend on the entrepreneur's residential history, but rather on their personality and innovation skills (see also Brulin 2007).

Nature, Living Heritage, and Local Stories as Resources

Local resources are utilised in the development of tourism. History and traces of the past, such as traditions and old buildings, are used as cultural capital (Grahn 2017). For the entrepreneurs in Merikarvia, the locality is important. Nature, the sea, the river, local fish, and fishing are strongly connected to the locality, and to the identity of the community. I asked the entrepreneurs what the place means to them and whether they could do business somewhere else. In many answers, the uniqueness of the area, and especially its maritime nature, was highlighted. Entrepreneurs operate in very different places, and changing the place itself is not necessarily a threshold issue. However, the essential thing is that they want to continue their activities in a rural area like Merikarvia, near the sea (Interviews 1–5).

Narratives play a key role in the development of tourism. The stories can be related to the history of the place, architecture and art, agriculture and food culture, industry, natural history, or religion. The creation of spatial narratives and imagined places is as essential as the physical environment itself. The interpretation of stories creates experiences. A good and high-quality interpretation may even increase the interest of the destination and attract the visitor to return (Meethan 2001; Smith 2015; Timothy 2011). In recent years, the popularity of storification has grown in tourism development along with stories. Storification is more than telling stories. It is about comprehensively shaping the service and products into stories. Storification aims to produce memorable and selling customer experiences. It supports productisation and differentiation, and brings meaning to services (Kalliomäki 2014).

Nature-based entrepreneurship has a strong foothold in rural tourism. In Finland, people's relationship with forests, and civil rights connected with this, have been in the National Inventory of Living Heritage since 2017 (Elävä perintö 2021). People's relationship with forests is often referred to as psychological ownership, which involves the thought or feeling that the object of ownership, the forest, 'belongs to me'. Psychological ownership is related to traditional ways of using nature, everyone's rights, and communality, and it does not require objective, that is, legal ownership (Matilainen 2019). The idea of 'psychological ownership' contains certain connections with cultural heritage. Participation in cultural heritage is independent of who owns the cultural heritage object. Cultural heritage belongs to all those who use it as evidence of history in the continuum in which they place themselves (Sivula 2009; Grahn 2014).

The entrepreneurs interviewed utilise the local cultural heritage in different ways. Cultural heritage utilised to produce products and services can be tangible or intangible. It can also be living heritage. The nature tourism entrepreneur who organises nature and canoeing trips utilises the nature of the nearby forests and the river Merikarvianjoki (see Figure 8.1). Everyone's rights are familiar and important to them, and psychological ownership comes out strongly in their thoughts as well: 'I feel that these forests are all mine.' As an entrepreneur, however, it is important for them to remember whose forests they are moving and working in. When they take groups to someone else´s forest, they always ask the landowner for permission to walk or to make a fire (Interview 1).

The nature tourism entrepreneur (Interview 1) runs his business in a small village of Alakylä, which itself has a long and colourful history.

Figure 8.1 Stone fences represent typical local cultural heritage of Merikarvia, and stories about stone fences are utilised on forest trips.

Source: Photograph by Maarit Grahn (2020).

A road called King's Road (*Kuninkaantie*) runs through the village. The road has got its name from the Swedish king Gustav II Adolf's visit to the village in 1614. Along the road there is also a history board made by the villagers, which tells about the history of the place. One of the entrepreneur's holiday cottages, located by the King's Road, is called *Kuninkaanmaja* (The King's Hut).

In producing experiences, the nature tourism entrepreneur utilises local stories. For example, there is a big rock, called *Nimikallio*, where local residents once used to gather. There are several engravings of initials and years made on the rock as a memory of this era. In the spring of 2020, the entrepreneur and several other active residents of village cleaned the rock from the layer of lichen and moss that covered it, so that the old carvings could be revealed. The oldest engraving that was found dates back to 1899. The entrepreneur also made signposts for the path leading to the rock. Nearby the rock there is a guest book where hikers can write their greetings, and a campfire place for a snack break (Interview 1).

The Merikarvianjoki river is one of the most popular fishing destinations in Southern Finland. The river is also particularly important to the nature tourism entrepreneur, because he has participated in a development project around the river. As a result of the project, many rapids and streams have been provided with resting places for fishermen and hikers, with sheds, fire pits, and toilets. According to the entrepreneur, one of the great advantages of the river is that it enables a rather long tourism and fishing season, as the river is accessible for at least three quarters of the year (Merikarvianjoki 2022; Interview 1).

Along with the river, maritime cultural heritage is an important element in building local identity and an image of the coastal region, as well as a tourism resource. In the marine environment, natural heritage and human achievements are closely intertwined. In the Oura archipelago, located in the Bothnian Sea National Park, maritime cultural heritage can be utilised in many different ways. The Oura-islet with its buildings, fauna, the visual landscape of the archipelago, the soundscape, and the history of the archipelago are the raw materials of experiences (Hankonen 2009).

The Oura archipelago is located in view of Merikarvia, about seven kilometres from the mainland. The old pilot station at Oura-islet was used by the Coast Guard until 1999. When the Coast Guard was moved from Merikarvia to Pori Kallo, the islet became the property of the municipality of Merikarvia (Sivula et al. 2017). At the beginning of the twenty-first century, the municipality began to develop Oura-islet for recreation and tourism. Since then, the two skippers have taken care of tourist transportations to the islet. The place interests both tourists and local people (Grahn 2022).

The knowledge gained as fishermen is a valuable resource for the skippers, as local knowledge is an absolute prerequisite for moving around the barren archipelago. The skippers also know a lot about the life of the Oura-islet, because they both have had pilots in their family histories. The skippers tell stories according to the customers' interests. There is not always a need for stories. Some of the passengers may come to the islet just to enjoy the scenery and nature of the Bothnian Sea. Sometimes customers may be interested in changes of the landscape, such as the effect of windmills or seal and cormorant populations (Interviews 3 and 4):

> It's the islet's history that people usually go there for. But the trip itself from the port to the islet is also interesting. Many people want to know the names of some shoals. 'Who does that belong to and what has been done there?' There are many interesting places, for example in Viitameri [sea bay] there are places where big ships were loaded in the past.
> (Interview 4)

Two of the local companies utilise especially rural cultural heritage and elements of country life in their business activities. Rural life, buildings, old work methods, beliefs, animals, and nature are the ingredients which the local farm tourism entrepreneurs have utilised over 30 years to produce nature and rural experiences for tourists. The cultural heritage is commercialised in the blink of an eye, so to speak. Locality and the traditional rural way of life are the elements that are utilised in producing tourism products.

Domestic animals (e.g., cows, sheep, horses, chickens, goats, and pigs) have an integral role in creating rural experiences (Interview 5):

> Everything local and peasant and all that kind of stuff has been our thing since the beginning. [...] So, I often talk about the recipe when I present our activities. And one part of the recipe is the locality and peasant life.
>
> (Interview 5)

The farm tourism entrepreneurs have utilised storification in building tourism services for several years. In storification, they have utilised their amateur theatre background. Storification has also been used to brand the family business. In 2008, the company began to emphasise the value of milking and the encounter with the cow. The generational change that took place in 2020 was also wrapped up in a story where the old host couple (i.e., the founders of the company) handed over responsibility for the business and the farm's main herd to the new farm manager (Interview 5).

The other entrepreneur whose activities are strongly intertwined with the rural cultural heritage is the local horse breeder. Horses are the central resource of the business, and at the same time, the breeder maintains the know-how and tradition related to Finnish horses (Interview 2). The Finnish horse is the only original horse breed developed in Finland. Due to its

Figure 8.2 Finnish horses and the power of nature are utilised in producing rural experiences.

Source: Photograph by Maarit Grahn (2020).

long history, the Finnish horse has a strong cultural and symbolic value, and knowledge related to Finnish horses has been in National Inventory of Living Heritage since 2017 (Elävä perintö 2021).

Along with the horse breeding work, nature and the power of nature, and the relationship between human and horse are strongly present in the operation of the horse farm (see Figure 8.2). Local stories related to places are used during horse-riding trips to local terrain. The entrepreneur utilises old stories she has heard from the local residents:

> We have a natural spring near here, from which all the villagers of Peippu used to get water. We organise riding trips to the spring. We ride there and drink spring water and do all kinds of things like that [...] We tell stories about these places, and what has been here before. That is kind of interesting.
>
> (Interview 2)

Rural Entrepreneurship and Future-Sustainable Development

The cultural and natural environment are the most important resources of tourism, and sustainable tourism can ensure the preservation of these resources. Sustainable development and sustainable tourism can be viewed from the perspective of economic and socio-cultural sustainability in addition to ecological sustainability (Ympäristöministeriö 2024). In sustainable tourism, tourism is developed responsibly without harming nature and culture, and promoting the well-being of local communities. The choices of sustainable tourism also involve new types of mobility and transport solutions and the promotion of equality, such as the development of accessibility in tourism services (Satakuntaliitto 2024).

In socio-culturally sustainable tourism, the relationship between tourists and the destination community is based on mutual respect, and tourism does not exploit the local population or damage the culture or the nature of the destination area. The development from the point of view of the local community is a basic requirement for sustainable development, to avoid conflicts of interest that can arise if the development of tourism is not based on the interests of the community. This seems to be true especially in rural areas due to the symbiotic relationship between the local community and the rural environment. The local community must have the opportunity to influence the development of their own area, and the development must be meaningful from their point of view (Hakkarainen 2017; Kugapi et al. 2019; Lordkipanidze et al. 2005; Wilenius 2004).

In rural areas, small businesses can have a favourable influence on the economic, socio-cultural, and ecological development of the region. They can strengthen local culture and identity, diversify tourism activities, and keep the rural population in the area. Thanks to the small size of the business, the environmental impact is also often minimal (Lordkipanidze et al. 2005).

Figure 8.3 The skippers' boats at Oura-islet. Ecological aspects have been taken into account when purchasing boats. The boats are fairly new; they have low effluent, and fuel consumption is minimised. They are also quiet in operation.

Source: Photograph by Maarit Grahn (2020).

In my study, the local entrepreneurs understood sustainable development primarily from the ecological point of view (see also Blinnikka 2012). The interviews highlighted the practical measures and choices that entrepreneurs make to promote sustainable development in their daily operations (see Figure 8.3). For instance, the entrepreneurs mentioned recycling and the utilisation of recycled materials. In addition to recycling, ecological thinking is emphasised in energy management (e.g., use of solar energy), forest management, farming, and consideration of the environment and nature (Interviews 1–5).

Socio-cultural sustainability as a concept was relatively unfamiliar to the entrepreneurs. However, socio-culturally sustainable operating methods are present in their activities. This came up 'between the lines' when I talked with the entrepreneurs about co-operation and values. Tourism and business activities are developed based on the local community's own interests, community oriented, respecting their own culture and cultural heritage (Interviews 1–5):

> I think it's really important in a small place like this for entrepreneurs to co-operate, to buy as much as possible from other local entrepreneurs. As we have done ourselves. And the locals have also been involved in the making of our horse-riding arena.
>
> (Interview 2)

The successful operation of small rural companies is based on local co-operation and networks (Summatavet and Raudsaar 2015). The co-operation also promotes and strengthens socio-culturally sustainable development. The entrepreneurial atmosphere is also crucial, because it creates the basic spirit in which entrepreneurs operate. The public sector also has an important role as an actor supporting and enabling entrepreneurship and co-operation networks (Lordkipanidze et al. 2005). Co-operation between entrepreneurs can be formal, but it is often informal and invisible (Tervo-Kankare 2019). According to the interviewees, the entrepreneurial atmosphere in Merikarvia is good. In addition to formal co-operation, local entrepreneurs do a lot of invisible co-operation: various services and raw materials are procured from neighbouring entrepreneurs (Interviews 1–5):

> We believe that tourism is the prerequisite for the entire municipality's independence and survival. And the cooperation of local entrepreneurs is very important, because there are no big companies here, just small businesses.
>
> (Interview 5)

Socio-culturally sustainable development includes nurturing and passing on the culture and cultural heritage of the target area. When refining cultural heritage into tourism products or services, entrepreneurs participate in the transmission, maintenance, and building of local identity and cultural heritage (see, e.g., Brulin 2007; Ilmonen 2016). According to my research, entrepreneurs consider themselves to be some kind of mediators: with their own contribution, they maintain interest in heritage. The entrepreneur sees an opportunity in cultural heritage, makes a choice, and works to bring it to life, to pass it on, whether it is the production of tourism services or other business activities for which cultural heritage serves as a resource. On the other hand, the choice also comes with the responsibility of which elements are chosen and how they are worked on, so that the legacy is preserved for the next generations. The interviews also highlighted the fact that the entrepreneur does not have the role of cultural heritage collector. Tasks related to storage and recording belong to the cultural heritage organisations. The role of an entrepreneur is to create experiences and make people see places, buildings, and things in a new way (Interviews 1–5; see also Brulin 2007).

Responsible and future-sustainable business and tourism also include the ability and courage to renew. Future sustainability means that companies and entrepreneurs have the necessary operating conditions to face and respond to the challenges of the future. This includes innovation development and strategic planning (Ketonen-Oksi 2021).

Digitalisation will bring plenty of opportunities in the future, but it also sets requirements. With the help of developing technology, tourism services can be offered in a more versatile and sustainable manner than before. In addition, tourism infrastructure can be utilised, and destinations can be managed more efficiently than before. The rapidly growing demand for sustainable tourism services must be taken into account in tourism companies, and with the help of digitalisation it is possible to better meet the demand and the needs of consumers. However, in order to utilise digitalisation, companies need support in digital skills. Small- and medium-sized companies operating in the tourism industry should have the same opportunities to utilise digitalisation so that they can compete in the global market (Työ- ja elinkeinoministeriö 2021).

Aspects related to digitalisation were also highlighted in the interviews. The functioning of network infrastructure in a rural area is seen as a primary condition for the future. Attention should also be paid to digital skills and training. The entrepreneurs use digital services in varying ways. Some of them use digital reservation, ticket sales, and payment systems, and the vast majority of entrepreneurs use social media applications (Facebook or Instagram) in addition to websites. On the other hand, the challenges related to time management came up in the interviews. Especially for sole proprietors it is sometimes difficult to find time to update social media (Interviews 1–5).

During the COVID-19 epidemic, a certain leap in the use of digital services was taken in Merikarvia. In the summer of 2021, an equipment rental company and a local food store, *Krookka Rent & Go Out*, were completed in the area of Krookka's guest marina. It is built from recycled containers, and it offers services with a new kind of concept. The customers handle purchases or rentals in advance virtually and pick up the products from the container with the code they received when ordering. For example, one can rent snowshoes, SUP boards, or fat bikes from the rental company. The container rental company was implemented by the municipality of Merikarvia, and local entrepreneurs are responsible for the rental and sales activities. In connection with the container rental company, a new company, *Krookan Plassi*, was also established. In addition to cafe services, this new company is responsible for the maintenance and development of guest marina services.[2]

One of the challenges for future-sustainable development is climate change. On the west coast of Finland, the winters are usually quite mild, which is why the area is not as popular as a winter tourist destination as northern Finland. But the area has its own special features, such as ice fishing and moving around on the sea ice. In recent years, climate change has begun to affect coastal winters and activities of tourism entrepreneurs. The last winters have been quite warm, and the sea has not frozen every winter. Especially for traditional ice fishing, the situation has been threatening. On the other hand, threats related to global warming can also be turned into opportunities by developing new methods of operation. For example, in the last few years, floating in rescue suits has become a popular winter activity (Merikarvia 2022). On the contrary, there have also been exceptional winters that offer surprising opportunities for the development of winter tourism. For instance, in the winter of 2022, the sea froze early and the ice cover remained for almost four months. This made it possible to get an ice rink for ice skating and traditional kick sledding on the ice of Krookka-bay in co-operation with the municipality, the local tourism association, and local entrepreneurs (see Figure 8.4).

Figure 8.4 In the winter of 2022, several local entrepreneurs participated in the ice rink project by sponsoring traditional kick sleds, which the public could use for free.

Source: Photograph by Maarit Grahn (2022).

Rural Vitality from Small Businesses

Nature is an essential part of rural cultural heritage and identity. It is also a key resource for the tourism business. Cultural heritage and nature are closely combined in the activities of the entrepreneurs in Merikarvia. Nature, river, and marine life of the coastal region are resources that can be used to create interesting content for tourism activities. Local nature entrepreneurship is small-scale tourism based on local resources.

Telling stories is an important part of the productisation of rural heritage and customer encounters. Entrepreneurs use storification in a rather variable way in their operations. The entrepreneurs' own starting points and the business affect whether storification is suitable for the activity and how much it can be used. Some entrepreneurs also feel that stories could be used more, and that their own storytelling skills would require polishing and practice.

Along with other local actors the entrepreneurs have an important role in cultural heritage processes and in advancing sustainable development. Entrepreneurs also have a great responsibility for which cultural heritage elements are chosen and how they are worked on so that the cultural heritage is preserved for the next generations. The research also revealed that not all entrepreneurs have necessarily thought about their role as mediators of cultural heritage, and in some entrepreneurs the research awakened new thoughts about their role in local cultural heritage processes.

According to the research, entrepreneurship based on cultural heritage and local identity are mutually reinforcing factors. While entrepreneurs draw from elements important to local identity, they maintain and renew cultural heritage and strengthen local identity through their activities. Entrepreneurs can also be viewed as a certain kind of ambassadors. They have an important role in building the image of the place.

Local tourism in Merikarvia is quite small-scale. Entrepreneurs working in tourism are mainly members of the local community, and socio-culturally sustainable ways of working seem to be self-evident for them (Olsen and Kugapi 2021). My research confirms the results of previous studies in that the promotion of entrepreneurship utilising local cultural and natural heritage can bring many economic, socio-cultural, and environmental benefits. In the future, the local tourism industry should be developed sustainably, based on the local community's own interests and respecting its own cultural heritage.

Utilisation of various online services and marketing will be a prerequisite for success in the future tourism market. Co-operation for the development of digital services may be the answer to the needs of small companies or sole proprietors, and the local entrepreneurs have already co-operated with the municipality of Merikarvia to develop electronic services and online shopping.

The companies involved in the research are typical small companies engaged in tourism in rural areas. One advantage of small companies is flexibility. This was especially visible in the research year 2020 during the restrictions brought by the COVID-19 epidemic. Entrepreneurs were required to be particularly innovative and flexible when they had to quickly think of new ways of operating for the production of tourism services and customer encounters.

Rural tourism gets its vitality from small businesses. With the promotion of business activities, the number of local tourism companies will increase, and existing companies can also become stronger. New business activities bring more jobs, business life becomes more diverse, and the quality of life improves. In addition, local culture and identity are strengthened when local services and products are supported and the local community is involved in business activities. At the same time, awareness of the value of cultural heritage increases, and thus the protection and preservation of rural cultural heritage and natural heritage can be better guaranteed. The activities of small companies can ensure the protection of natural diversity, minimise environmental impacts, and produce sustainable tourism products and services (Lordkipanidze et al. 2005). Popularity of nature and local tourism is constantly growing. This creates good conditions for the development of tourism and for the responsible use of rural cultural heritage, nature, and the marine environment.

Notes

1 For my research, I chose a wide variety of entrepreneurs because one of my goals was to form a broad picture of how cultural heritage is used as a resource and what types of entrepreneurs work with cultural heritage. I contacted twelve entrepreneurs, and finally, ten entrepreneurs or actors who work like entrepreneurs participated in my study. They all represent small businesses with one to five people working on a permanent basis. Some of them are sole proprietors, some are small family businesses. Some of the entrepreneurs have been operating for a year or two, some have been operating for several decades. The following entrepreneurs participated to the research: one nature tourism entrepreneur, one entrepreneur and two family businesses who run cafe or restaurant operations in industrial cultural environments, one entrepreneur who engages in breeding and riding Finnish horses and organizes cross-country hikes for riders, two former fishermen who manage boat transport to the Oura archipelago, two blacksmiths who have just started their business, and one family business that has been engaged in farm tourism for 30 years. All the interviews were taped and transcribed for the analysis. The permission of the interviewees has been obtained for the use of the interview material. The material has been in the researcher's possession, and after the research was completed, the interview material has been deleted by the researcher.

2 The municipality of Merikarvia received funding for the Krookka Rent & Go Out project from the Regional Council of Satakunta. The activities of Krookan Plassi are based on the principles of sustainable development. Krookan Plassi was not involved in this study as a company, but is connected to the sustainable development of tourism at the local level.

References

Blinnikka, Petra. 2012. "Kestävä matkailu maaseutumatkailuyrittäjien silmin. Raportti keväällä 2011 toteutetusta kartoituksesta Keski-Suomen, Pirkanmaan, Pohjanmaan ja Etelä-Pohjanmaan alueilla." Jyväskylän ammattikorkeakoulun raportteja nro 19. Jyväskyl: Jyväskylän ammattikorkeakoulu.

Bowitz, Einar and Karin Ibenholt. 2009. "Economic Impacts of Cultural Heritage – Research and Perspectives." *Journal of Cultural Heritage* 10: 1–8. https://www.sciencedirect.com/science/article/abs/pii/S1296207408001544

Brulin, Göran. 2007. "Framtida näring – identitet som drivkraft." *Kulturarv och entreprenörskap. Dokumentation av konferens i* Jönköping 10–11 Maj, 2007. Stockholm: Riksantikvarieämbetet. https://www.raa.se/publicerat/9789172095120.pdf

Elävä perintö. Wiki – inventory for living heritage. https://wiki.aineetonkulttuuriperinto.fi/

Gadamer, Hans-Georg. 2004. *Hermeneutiikka: ymmärtäminen tieteissä ja filosofiassa.* Tampere: Vastapaino.

Grahn, Maarit. 2014. *Perheyhtiö ja paikallisuus. A. Ahlström Osakeyhtiön historian perintö Noormarkussa.* Turun yliopiston julkaisuja. Sarja C. Osa 374. Turku: University of Turku. https://www.utupub.fi/bitstream/handle/10024/94239/AnnalesC374.pdf?sequence=2&isAllowed=y

Grahn, Maarit. 2017. "Time Travels and Memory Sites. Cultural Heritage of a Family Enterprise as Business Resource." *Proceedings of the Heritage. Tourism and Hospitality International Conference HTHIC, Pori, Finland, 27–29 September, 2017.* Publication of Turku School of Economics, Pori Unit, A55/2017, 77–88. Turku:

University of Turku. https://www.utupub.fi/bitstream/handle/10024/144335/Proceedings%20Final.pdf?sequence=1&isAllowed=y

Grahn, Maarit. 2021. "Kulttuuriperintöyrittäjyys maaseudun voimavarana." *Matkailututkimus* 1/2021: 7–25. Oulu: Suomen matkailututkimuksen seura. https://journal.fi/matkailututkimus/article/view/107796/64515

Grahn, Maarit. 2022. *Kulttuuriperintöyrittäjyys, elinvoimaisuus ja tulevaisuuskestävä ympäristö – Case Merikarvia*. Kulttuurituotannon ja maisemantutkimuksen julkaisut 60. Turku: University of Turku. https://www.utupub.fi/bitstream/handle/10024/153919/Kulttuuriperint%c3%b6yritt%c3%a4jyys%20elinvoimaisuus%20-%20Case%20Merikarvia.pdf?sequence=1&isAllowed=y

Hakkarainen, Maria. 2017. *Matkailutyön ehdot syrjäisessä kylässä*. Acta Universitatis Lapponiensis 357. Rovaniemi: University of Lapland. https://lauda.ulapland.fi/bitstream/handle/10024/63004/Hakkarainen_Maria_ActaE_224_pdfA.pdf?sequence=1&isAllowed=y

Hankonen, Ilona. 2009. *Luonto ja paikalliskulttuuri matkailun voimavaroina Merikarvialla: toiveet, haasteet, mahdollisuudet ja visiot matkailun toimijoiden silmin*. Turku: University of Turku.

Hirvonen, Sini. 2017. "Vapaaehtoiset hoitavat kulttuuriympäristöä – ja huolehtivat samalla hyvinvoinnistamme." In *Elossa. Luonto ja elävä kulttuuriperintö*, edited by Aura Kivilaakso and Leena Marsio, 23–33. Helsinki: Museovirasto. https://www.aineetonkulttuuriperinto.fi/assets/elossa-Luonto-ja-elava-kulttuuriperinto.pdf

Ilmonen, Kari. 2016. *Kulttuuriperintö maaseudun matkailuyritysten tuotteissa ja palveluissa*. Kokkola: Jyväskylän yliopisto, Kokkolan yliopistokeskus Chydenius. https://jyx.jyu.fi/handle/123456789/49408

Kaipainen, Jouni. 2015. "Maaseudun kulttuuriperinnön hyödyntäminen." *Maaseudun uusi aika* 3: 65–74. https://www.mua-lehti.fi/arkisto/3_15/kaipainen.pdf

Kalliomäki, Anne. 2014. *Tarinallistaminen. Palvelukokemuksen punainen lanka*. Helsinki: Talentum.

Ketonen-Oksi, Sanna. 2021. *Transforming how we think about futures*. Accessed 21 October 2024. https://ketonenoksi.com/

Korsgaard, Steffen, Sabine Müller and Hanne Wittorff Tanvig. 2015. "Rural Entrepreneurship or Entrepreneurship in the Rural – Between Place and Space." *International Journal of Entrepreneurial Behavior & Research* 21 (1): 5–26. https://www.emeraldinsight.com/doi/full/10.1108/IJEBR-11-2013-0205

Kugapi, Outi, Emily Höckert, Monika Lüthje, Nuccio Mazzullo and Ritva Saari. 2019. *Kohti kulttuurisensitiivistä matkailua. Suomen Lappi*. Matkailualan tutkimus- ja koulutusinstituutin julkaisuja. Rovaniemi: Lapin korkeakoulukonserni. https://lauda.ulapland.fi/bitstream/handle/10024/64013/Kohti_kulttuurisensitiivist%c3%a4_matkailua.pdf?sequence=1&isAllowed=y

Lithander, Jörgen. 2002. "Samhällsentreprenörerna i Naestie – lokal utveckling i fjällmiljö." In *Kooperation, Ideellt arbete & lokal ekonomisk utveckling. Kooperativ Årsbok 2002*, edited by Ann-Mari Sätre Åhlander, 187–216. Stockholm: Föreningen Kooperativa Studier. https://svenskkooperation.se/wp-content/uploads/2016/05/ka2002.pdf

Lordkipanidze, Maia, Han Brezet and Mikael Backman. 2005. "The Entrepreneurship Factor in Sustainable Tourism Development." *Journal of Cleaner Production* 12: 787–798. Elsevier. https://www.sciencedirect.com/science/article/abs/pii/S0959652604000605

Malkamäki, Sisko and Antti Tulila. 2012. "Alueelliset vahvuudet ja resurssit maaseudun kehittämisen lähtökohtana. Case: Tervon kunnan kalatietokeskus." In *Uudistuva yrittäjyys maaseudulla. Esseitä ja esimerkkejä. N:o 187/2012*, edited by Tarja

Niemelä, 33–44. Jyväskylä: Jyväskylän yliopiston kauppakorkeakoulu. https://jyx.jyu.fi/bitstream/handle/123456789/38275/978-951-39-4793-4.pdf?sequence=1

Matilainen, Anne. 2019. "Feelings of psychological ownership towards private forests." Ruralia Institute Publications 36. Helsinki: University of Helsinki. https://helda.helsinki.fi/bitstream/ handle/10138/300433/FEELINGSO.pdf?sequence=1&isAllowed=y.

Meethan, Kevin. 2001. *Tourism in Global Society. Place, Culture, Consumption.* New York: Palgrave.

Merikarvia. Kummallisen kodikas 2022. Accessed 21 October 2024. https://merikarvia.fi/

Merikarvianjoki. Niin lähellä, mutta niin kaukana arjesta! 2022. Accessed 21 October 2024. https://www.merikarvianjoki.fi/

Metsähallitus. Luontoon.fi. Selkämeren kansallispuisto. Accessed 21 October 2024. https://www.luontoon.fi/

Molin, Torkel. 2007. "Att varugöra ett kulturarv." In *Industrihistoriska kulturarv i regional utveckling – sociala och ekonomiska aspekter. Konferens för kunskapsutveckling och nätverksbyggnad. Rapport från Riksantikvarieämbetet 2007:4*, edited by Torkel Molin and Martin Paju, 18–22. Stockholm: Riksantikvarieämbetet. https://www.raa.se/publicerat/9789172094680.pdf

Olsen, Kjell O. and Outi Kugapi. 2021. "On Tourism Guidelines and Roadmaps in the Arctic. Arctisen. Culturally Sensitive Tourism in the Arctic." *ARCTISEN* Blog, 12 March 2021. https://blogi.eoppimispalvelut.fi/arctisen/arctisen-blog/

Satakuntaliitto. Matkailun kasvuohjelma 2030. Accessed 21 October 2024. https://satakunta.fi/aluekehitys/strategiat-ja-ohjelmat/muita-keskeisia-strategioita-ja-ohjelmia/matkailun-kasvuohjelma/

Sievänen, Tuija. 2017. "Luonnossa virkistäytyminen pitää yllä yhteyttä luontoon ja kulttuuriperintöön." In *Elossa. Luonto ja elävä kulttuuriperintö*, edited by Aura Kivilaakso and Leena Marsio, 16–22. Helsinki: Museovirasto. https://www.aineetonkulttuuriperinto.fi/assets/elossa-Luonto-ja-elava-kulttuuriperinto.pdf

Siivonen, Katriina. 2019. "Elävä perintö ja kestävä kehitys." *Arvokas luonto. Aineettoman kulttuuriperinnön luontoringin seminaari 3.9.2019.* Helsinki: Kansallismuseo. https://www.aineetonkulttuuriperinto.fi/assets/Elava-perinto-ja-kestava-kehitys-Katriina-Siivonen.pdf

Sivula, Anna. 2009. "Pieniä mutaatioita. Menneen jäljet ja historia identiteetin rakennusmateriaalina." In *Matkalla. Juhlakirja Outi Tuomi-Nikulalle hänen täyttäessään 60 vuotta 19.10.2009*, edited by Helena Ruotsala, Petri Saarikoski and Maija Santikko, 261–271. Kulttuurituotannon ja maisemantutkimuksen laitoksen julkaisuja XVIII. Turku: University of Turku.

Sivula, Anna, Anni Ruohomäki and Riina Haanpää, ed. 2017. *Kuunarin kryssi. Merikarvialaisten historiaa 1860-luvulta 2000-luvulle.* Merikarvia: Merikarvian kunta.

Smith, Stephen. 2015. "A Sense of Place: Place, Culture and Tourism." *Tourism Recreation Research* 40 (2). London and New York: Routledge. https://www.tandfonline.com/doi/full/10.1080/02508281.2015.1049814

Summatavet, Kärt and Mervi Raudsaar. 2015. "Cultural heritage and Entrepreneurship – Inspiration for Novel Ventures Creation." *Journal of Enterprising Communities: People and Places in the Global Economy* 9 (1): 31–44. https://www.emerald.com/insight/content/doi/10.1108/JEC-03-2013-0010/full/html

Tervo-Kankare, Kaarina. 2019. "Entrepreneurship in Nature-based Tourism under a Changing Climate." *Current Issues in Tourism* 22 (11): 1380–1392. https://www.tandfonline.com/doi/pdf/10.1080/13683500.2018.1439457?casa_token=ZIE8

FmDeX6UAAAAA:pfTR1WvHX1C-85RHadKGAoILqpkZC-nO6TfaQ6EyUlq7qkOTcdSYahm4I6X8qTM7jrEHYwZ7IQSMDOGs1w

Timothy, Dallen J. 2011. *Cultural Heritage and Tourism. An Introduction.* Bristol: Channel View Publications.

Työ – ja elinkeinoministeriö 2021. Accessed 21 October 2024. https://tem.fi/etusivu

Wilenius, Markku. 2004. *Luovaan talouteen. Kulttuuriosaaminen tulevaisuuden voimavarana.* Helsinki: Edita.

Yachin, Jonathan Moshe. 2019. "The Entrepreneur-opportunity Nexus: Discovering the Forces That Promote Product Innovations in Rural Micro-Tourism Firms." *Scandinavian Journal of Hospitality and Tourism* 19 (1): 47–65. https://www.tandfonline.com/doi/abs/10.1080/15022250.2017.1383936

Ympäristöministeriö. Mitä on kestävä kehitys? Accessed 21 October 2024. https://ym.fi/mita-on-kestava-kehitys

Zaman, Gheorghe. 2015. "Cultural Heritage Entrepreneurship (CHE) – Challenges and Difficulties." *Procedia – Social and Behavioral Sciences* 188: 3–15. Elsevier. https://www.sciencedirect.com/science/article/pii/S1877042815021266

9 The River Eurajoki

A Source of Local Natural and Cultural Heritage

Riina Haanpää, Laura Puolamäki, and Eeva Raike

The River Eurajoki has been the focus of the Degree Programme in Digital Culture, Landscape, and Cultural Heritage at the University of Turku since the early 2010s. Our research interest has been stimulated by the various environmental conflicts that the River Eurajoki has faced over the past years and decades – industrial pollution and emissions, nutrient loads, and waste-water run-offs have put pressure on the river, leading authorities to declare it to be in poor ecological condition. However, the status of the river has been a matter of concern for residents and entrepreneurs along its course, to the extent that dialogue between the local community and the authorities has not always been without problems. It is precisely in this lack of dialogue that the degree programme saw the potential for multidisciplinary and cultural-environmental studies to contribute to the debates surrounding the River Eurajoki. At the same time, we found ourselves pondering why the local ecological knowledge held by the community is rarely valued, for example, in formal environmental licensing processes.

In this chapter, we examine the interaction between the River Eurajoki and the local community. We first review the use and importance of the river, both in the daily lives of local people and in the pursuit of economic activities and industrial production. We then present the field studies we have carried out on the River Eurajoki and their objectives. Next, we turn our attention to local ecological knowledge and the cultural significance of the river that has emerged from the data collected in the field. We will discuss which values the river has carried and continues to carry, and how best to reach out to these values. Our aim is therefore to highlight how the local community's cultural relationship with the environment generates the qualitative data that would greatly benefit environmental decision-making processes.

The Importance of Human Activity – Past and Present of the River Eurajoki

The River Eurajoki runs through the municipalities of Eura and Eurajoki in the Lower Satakunta Region in the south-west of Finland. The river is approximately 52 km long, and the watershed covers an area of 1,336 km^2.

DOI: 10.4324/9781003727972-9

Today, the river looks quite modest, partially due to the land uplift process gradually reconstructing the landscape of the west coast of Finland. In earlier times, however, the river was an important thoroughfare and source of livelihood for the region (Anttila et al. 2022, 15).

The significance of the River Eurajoki for local inhabitants and entrepreneurs has been and remains manifold, although the occasional pollution of the river water, especially in the twentieth and twenty-first centuries, has forced inhabitants to periodically withdraw from its banks. Historically, the river has been accessible by boat in the summer and by skiing and horse-drawn sleigh in the winter when the ice cover has formed. At its best, navigation along the icy waterways has been almost unimpeded. During the melting season, crossing the river remained a necessity, so prior to the construction of bridges, efforts were made to find fords or wading places where the water was low, and transit was easy. Long poles or paved crossing points may also have been used to cross the river (Anttila et al. 2022, 9). Then, as the volume of land transport increased and the road network expanded, bridges were built to facilitate river crossings. It is not possible to date the first bridge over the River Eurajoki with certainty, but geometric maps of the region from as early as the 1690s show several early wooden bridges across the river. The oldest stone bridge over the river, built in 1832, is the double-arched bridge in the village of Panelia, which is still used for light traffic today (Anttila et al. 2022, 11).

Historically, the River Eurajoki provided drinking water for both people and their livestock, supplied irrigation, and acted as both the source and location of work and leisure. The river was the most convenient place for women to do their laundry: laundry pads were brought to the riverbank, and after washing, the laundry was rinsed in the flowing water of the river. The Finnish custom of washing hand-made carpets in the river every summer has been practiced for decades. In winter, the river was a source of ice, which was taken and stored in sawdust throughout the summer for households and the Panelia village dairy, for example.

The River Eurajoki has a rich fish population, although the quantity and variety have varied over the decades. Atlantic salmon spawned in the river prior to the construction of the Pappilankoski power plant in 1932, making it a significant fishing river from early on. Skimmer fish have also been caught in the river, at least in the Tiironkoski and Nolponkoski areas (Heikki Falttu 23.4.2014), and were fished professionally until the late 1940s. Other fish have also been present in the river at different times. Nowadays, children in the area can identify multiple fish present in the river, such as pike and perch (Anttila et al. 2022, 92).

As the Atlantic salmon disappeared from the river and the water quality deteriorated, the freshwater pearl mussel population began to decline and gradually disappear from the river fauna. The requirements for mussel habitats are strict, especially during the larval and juvenile stages, requiring neutral and acidic water, good flow conditions, and a stock of trout or salmon as intermediate hosts for the larvae. Freshwater pearl mussels have been caught in the River Eurajoki for their precious pearls. Alongside pearl fishing, the mussel population was depleted by the dredging, damming, and regulation of the river. Industry in the upstream slowly polluted the water and destroyed not only the mussel population but also a wide range of other biota (Anttila et al. 2022, 104).

Salmon and mussel were almost certainly historically important for the River Eurajoki, as several maps from the seventeenth and eighteenth centuries mark their presence in the river. Liinmaa Castle, which stands today on the mainland between the River Eurajoki and the River Lapinjoki, is another indication of the importance of the area. In the thirteenth century, the castle was located on a small island at the mouth of the River Eurajoki, protected by a land wall (Bläuer 2019, 261, 271–272; see also Uotila et al. 2011; Anttila et al. 2022, 32). Access to and utilisation of the river may have been guarded from Liinmaa Castle. It is likely that the pearls and the salmon from the river were so valuable that their gathering had to be controlled. Indeed, the River Eurajoki served as a gateway from the sea to the prosperous areas of Eura and Kauttua in the river mouth, which were inhabited in the prehistoric and early medieval period.

The River Eurajoki is deeply embedded in the daily lives of the people living along its banks. Klaus Roth (1997, 26) has noted how rivers can act as barriers to administrative or military activity, for example, but more often they have been a lifeline and 'bridge' between different cultures. Whether natural, economic, or cultural, rivers are also dispersal routes. Since time immemorial, the river has been used for navigation, fishing, and pearling. The flowing water of the river has also been used as a source of power, and the 1875 mill taxation census mentions eight mills on the Eura, where the inhabitants of the region used to grind their grain. Sawmills were also built along the river, and one of Finland's oldest known sawmills was located in the village of Irjanne, built at the Saharinkoski rapids in the 1630s (Anttila et al. 2022, 71–73, 78). Through mills and sawmills, the river indirectly provided a livelihood for the people dwelling along the river.

Until the end of the nineteenth century, the water of the River Eurajoki could still be used for economic and drinking purposes. Gradually, however, the quality of the water deteriorated with the advent of industrialisation, culminating in 1908 when the Kauttua wrapping paper mill of A. Ahlström Oy was established. The mill discharged paper pulp into the river, which turned the water into paper wastewater. With the threat of the mill's closure, an Imperial Senate decision in 1913 forced the wrapping paper mill to build a water treatment plant, which had a positive impact on the river's water quality in the 1920s and until the mid-1930s (Virtanen 2022, 41–42).

Urpo Heikkilä, who lived in the village of Eurajoki, recalls how in the mid-1930s the water became cloudy and began to change colour, after which the river water became clearer and the fish died (Heikkilä 2000). The water's clarity and the fish deaths were caused by A. Ahlström Oy, which used chlorine to bleach paper. The factory discharged the waste directly into the river, which once again contaminated the water. However, the factories along the river were not the only culprits for the pollution, as the River Eurajoki was commonly used as an open sewer in the area, as was the case with many rivers around the world, including Finland. For example, the Aura River, which flows through Turku, was at its worst state in the late nineteenth and early twentieth centuries, when it was used as a sewer for the city. The malodorous state of the river persisted until 1966, when the city's sewage treatment plant was completed (Valla 2019).

Finland's 1962 Water Act introduced the concept of *water protection*. Under the provisions of the Water Act, discharging sewage directly into the river was prohibited, and intermediate cesspools became mandatory. The Kauttua factory of A. Ahlström Oy was also required to build a water treatment plant, which significantly reduced discharges (Virtanen 2022, 43). Since the Act came into force in the 1960s, the load on the River Eurajoki has been monitored, and the river has sometimes fared better and sometimes worse. The most recent major deterioration in water quality was observed in 2012, when it was discovered that untreated wastewater from the JVP-Eura Oy wastewater treatment plant had been regularly discharged into the River Eurajoki because the plant did not have sufficient capacity to handle the influent load, especially in times of rain showers or spring floods.

In the same year, industrial plants and a wastewater treatment plant upstream also began to renew their environmental permits, which always apply for fixed terms only. The permit applications referred to the cultural and social values of the River Eurajoki, but only to the fact that the river had *little recreational value*. This reference infuriated the inhabitants of the river valley, who had spent their whole lives with the river and were now forced to retreat to its banks because of the poor water quality. A citizens' movement, the *River Eurajoki Working Group*, quickly sprang up around the activities of residents and entrepreneurs, who took an organised approach to defending the river. The citizens' movement began to call for stricter limits on the environmental permit of the wastewater treatment plant, and also for new measures to restore the river. The group also wanted to get involved in the preparatory process of the environmental permit, thus making the views and position of the local community heard by the authorities. Later, the movement was accompanied by the *Our Eurajoki* project, in co-operation with the University of Turku, the Pyhäjärvi Institute, and JVP-Eura Oy, which in turn aimed to improve the quality of the river's fresh water and the river habitat (see Figure 9.1).

Figure 9.1 The River Eurajoki at its best in the evening sun.

Source: Photograph by Tuomo Hurme (2000), used by permission.

The Importance of Local Ecological Knowledge – Field Studies along the River Eurajoki

The aim of the multidisciplinary project *Our Eurajoki* was to promote the sustainable use and protection of the River Eurajoki. There was a great need for this project, especially as environmental permit applications had led to a debate between the different parties which had become bogged down in arguments. The local community had only lay knowledge of the (cultural) significance of the river, as there was no research data available to support it in the discussions. In contrast, the decisions on applications for permits for industrial plants and sewage treatment plants were based only on data on the environmental state of the river, as part of the permitting process, which was in turn based on natural scientific monitoring. Thus, the need to understand the meaning given to the river by the local community, that is, residents and entrepreneurs, was not only topical, but also needed to be combined with the natural scientific data collected from the river, such as ecological, hydrological, and fisheries data. The joint project enabled this objective through the work of various actors. The Pyhäjärvi Institute monitored the river water quality with various continuous gauges, providing information on the river load and allowing better management of the environmental risks associated with the river. The Degree Programme in Digital Culture, Landscape, and Cultural Heritage aimed to determine what values the local community associated with its river. We hoped to bring to the fore the part of *local ecological knowledge* (LEK) that was not captured in the environmental planning processes.

The advantages of local ecological knowledge have been studied in recent years in forest management, sustainable urban planning, and aquaculture, among others. Such knowledge is relevant in attempts to understand any natural process, and especially in management objectives to fortify biodiversity and cultural relationships to nature. LEK refers to the knowledge, practices, and beliefs of local communities regarding ecological relationships (Yli-Pelkonen and Kohl 2005; Charnley et al. 2007; Aswani et al. 2018). This is an important part of local community (Aziz et al. 2022). A concept close to LEK is *traditional ecological knowledge* (TEK). Traditional ecological knowledge includes practices and beliefs that have been transferred between generations. TEK is usually included in indigenous knowledge. LEK can also be transferred between generations, and it can include traditions and beliefs, but it is thought to draw on scientific knowledge as well.

Initially, in 2013 and 2014, we carried out field research with students from the degree programme among residents of the villages of Panelia and Irjanne. We were specifically interested in the experiences, perceptions, and aspirations of the local community along the river, with the aim of making the cultural significance of the river visible and giving a voice to the locals (Fingerroos and Haanpää 2022, 469; Shopes 2002, 588). Our aim was also to observe the process of interpreting one's own habitat as a place lived and experienced. We therefore included not only interviews with local residents, which were integral to their experience of the habitat and thus created an environmental relationship, but also extensive literature on the history of the area, as well as photographs and maps.

Local residents and entrepreneurs[1] played an important role in informing us of the history and events of the area's past, as well as its current daily life and cultural environment. Through interviews,[2] we drew on their personal reflections and the cultural significance of the River Eurajoki (Haanpää 2014, 34; Hämeenaho and Koskinen-Koivisto 2014, 12; see also Fingerroos and Haanpää 2022, 470–472). Through the interviewees' lives, day-to-day experiences, leisure time, and hobbies, we attempted to determine what kind of place the river had been in their childhood, youth, and present, as well as what kind of places, stories, and memories were associated with the river (see Figure 9.2). For local residents and entrepreneurs, the cultural environment of the River Eurajoki was based on working, moving, and being in its spaces and places. In everyday life, for example, our sense of sight does not dominate our experience of the world, but as physical beings we experience the environment through all our senses and focus our attention on the objects that are central to our activities. The relationships between spaces, places, and things are indeed perceived in movement, in moving from one place to

Figure 9.2 Swimming, diving, jumping into the water, and other water activities were a big part of children's as well as adults' summer life.

Source: Photograph by Euran kotiseutuarkisto (1950s).

another, and in doing, and are thus always lived or experienced – we inhabit and live in them. The River Eurajoki is also such an environment, in the vicinity of which people had lived and continued to live, and in which they made their observations at different times. As bodily beings, the inhabitants of the riverside experienced their environment with all their senses, and the Eurajoki riverside was thus the cause of all the experiences that local people had had or could have had of the river (see Merleau-Ponty 1999 [1945], 81; 2002, 244):

> There was quite a big bay here before the river was dammed, and it was nine metres deep. There was a municipal bathing establishment here, and there was a municipal laundry dock here. People used to swim here a lot and wash their laundry. So, it was kind of a gathering place.
>
> (Seppo Aarikka 10 October 2020)[3]

We gathered information about the river's uses and livelihoods, fishing, carpet washing and laundry, logging, as well as industry with its sawmills, mills, and power stations. During the walking interviews, we surveyed the rapids and riverbanks, the neighbouring grazing areas, and the overgrown riverside and their impact on the landscape. We also asked the people living along the river about their feelings on the current state of the river and, above all, we wanted to listen to them. The fieldwork not only provided us with new perspectives to consider, but also enabled us to reach out and ascertain the interviewees' and the local community's own experience of the river's importance (Haanpää 2014, 32; see also Jokela 2021, 76; Ukkonen 2000, 14). In this way, the understanding of the River Eurajoki cultural values was brought to the attention of people outside the local community:

> The river, you know, is so familiar that you used to wash your laundry on the riverbank. Actually, it's a kind of summer day when you go to the riverbank. There was a big lunch bag, there was stuff on the horse's cart, there was a big pot and caps and white clothes, and especially men's clothes with flannel and linen, and it was almost a day there. You carried water from the river to the pot and then you washed it and then you were there and it was fun for the children.
>
> (Aino Lähteenmäki and Kaino Nurminen 23 April 2014)

In the interviews, photographs of life in the River Eurajoki in the old days served as excellent stimuli and refreshers of memory. As visual memories of the river's water status, the photographs told us what had been before and what was now. The photographs concretised the passage of time and also the crisis the river was facing. Images of the local community engaging in summer bathing days, washing carpets and laundry (see Figure 9.3), or grazing along the river with their animals not only evoked nostalgic memories, but also provoked opinions and helped field researchers, for example, to understand the emotional reactions to the current state of the river expressed by the interviewees (Kortelainen 2008, 65–67; Bagayoko and Tawah 2014, 168, 206).

In turn, we used old map material to explore changes in the riverbed, such as the spread of settlements along the riverbanks, the expansion of fields and meadows, and the drainage of various marshes and small lakes, all of which changed not only the landscape but also the ecological state of the river (see, e.g., Campbell 2021, 214–215, 217–218).[4] We used, for example, cadastral maps from the seventeenth century and maps of the land divisions, as well as watercourse maps and photographs, as aids in the interviews. These resources helped us to locate the information described by interviewees, in particular the place names and stories, by placing them on the map, and progress to detailing the history of the area and the farms that had been formed:

Interviewer: But here's this parcel map, can you see the land?

Interviewee: They are here then. Wait a minute, here goes this plot and this is number 15.

Interviewee: It's a farm estate, then?

Interviewee: Yes, it was divided into two in the nineteenth century, and the fields are right here on the riverbank. But this map then breaks off here.

(Jaakko Heiska 7 May 2013)

Figure 9.3 'When we went to riverbank, there was a carpet dock, which nearby farmers built every summer. Everybody got to use it for carpet and laundry washing and also swimming.' (Tapani Suomi 7 May 2013).

Source: Photograph by Jarmo Suomala (1958)/University of Turku, DMKT.

In all, almost 60 residents were able to share their views and experiences, as well as their memories of the riverside, during the public reminiscence evenings and personal interviews during the fieldwork. We mapped this local environmental knowledge to get to the heart of what was thought and talked about the state of the River Eurajoki – both past and present – and why and what was said (Abrams 2010, 1, 6–7). The aim of the interviews was therefore first and foremost to interpret the culture that was relevant to the locals and that had generated an active movement among the people of the River Eurajoki.[5]

The work of the *River Eurajoki Working Group*, which was formed among the local community of the Eurajoki arm of the river, also produced visible results. Thanks to the citizens' movement, for example, the permit conditions for industrial plants upstream were tightened in 2014, untreated wastewater discharged into the river was subjected to special monitoring, and in 2016, treated wastewater was finally sanitised before being discharged into the river. As a result of these measures, the River Eurajoki recovered and, as a concrete expression of this, the beach in the middle of the river was restored for the use of the local community. These results also prompted us researchers to start investigating whether the improvement in the ecological status of the river had led, for example, to an improvement in the environmental relationship between the children of the local community and the river.

The data collected in 2013 and 2014 were supplemented by field surveys in 2019 and 2020. This time, the focus was on children in the local community. Research workshops were held in two schools, in the villages of Panelia and Irjanne, where students shared their local ecological knowledge about the river with the researchers and collaborated to collect new data by interviewing adults in turn. When working with children, we considered how to ensure their voluntary participation in the study. In preparation for the workshops, we had agreed on the research arrangements with the schools, but as we approached the fieldwork phase, we wanted to ensure that the children's guardians and the children themselves were involved in the research. We sent a consent form and a data protection statement to each student's home for them to read and sign, and the form was signed by both the guardians and the children themselves. At the start of the workshops, each child was individually asked whether they wanted to participate in the study.

In the first phase of the workshops, the students described their own relationship with the River Eurajoki by sharing their own experiences of the river and by marking important places along the river on a map. In the second phase, the same students interviewed one or two generations of older riverside residents, repeating to them the same research questions they had answered themselves. These places along the riverbank were also plotted on the map. In this way, local knowledge about the river was integrated across generations:

> Behind the mill there are lots of rocks, and if you jump over them you can cross the river. You can hear birds singing and chirping. Behind the mill is a small island with a small stream running alongside it.
>
> (Eurajoki workshop at Panelia School, 4 September 2019, 12-year-old student)

In this way, the methods we had previously used in our fieldwork were augmented by participatory geographic information systems (PGIS) and citizen science.[6] We also continued to conduct research interviews, but in addition to recording them as audio files, we also videoed them and edited them into a mini-documentary.[7]

In the fieldwork with children, we specifically sought to understand the environmental relationship between children and the River Eurajoki today.

Previous fieldwork data from 2013 to 2014 recounted – as recalled by adults – the river relationship between children and their childhood from past decades. The time span was from the 1930s to the 1980s, after which the river relationship between children and their childhood was coloured by periods of environmental disasters and gradual degradation, during which the river had occasionally been closed for long periods, or in such poor condition that recreational use was unthinkable. In her research on the relationship between children and the cultural environment, Laura Puolamäki has noted how the River Eurajoki, which runs through the village of Panelia in the Eura municipality, has become disconnected from children's lives, because the ecological condition of the river has made it impossible for children to use this part of their environment in their everyday lives.[8]

The Importance of Environmental Awareness – Applying Local Ecological Knowledge

Although the 2013 and 2014 field studies aimed to gather information primarily on the cultural values of the River Eurajoki and its recreational use, the river's present and its ecological status also featured strongly in the interviews. In particular, the pollution of the river was a matter of concern, and local residents and entrepreneurs were aware of their environment and the 'persistence' of the poor state of the river to this day:

> The pollution is not something that happened in the last few years. The river was classified as permanently polluted sometime in the 1970s, claiming that there's nothing to be done about it. This tannery was pretty much dumping everything into the river back then.
>
> (Mikko Tolvi and Timo Tolvi 6 May 2013)

It is in this kind of collective discussion, such as the interviews conducted during the fieldwork, that the fruitfulness of human-based research data is formed. When the focus is on the local inhabitants' and entrepreneurs' own living environment and the threats or actions taken against it, and when the joint examination of the issue generates information that can be translated into visible results, it opens up new perspectives on the living environment. The joint study and the results that emerge from it bring together different types of information on the same site, the River Eurajoki. The results allow us to strive to highlight, articulate, and understand the cultural values of the River Eurajoki. They will also open up opportunities for dialogue between decision-makers and the local community, for example, on culturally responsible and sustainable business practices. Without cultural-environmental research, and the results it produces, dialogue between the various parties may only progress to a debate, as we first witnessed in the case of the River Eurajoki. The new, local ecological knowledge makes it possible to take into account the cultural values held by the local community, even if they are not required, for example, in environmental licensing procedures. This case study confirms earlier findings, that the best way to advance knowledge is by knowing together. This can be achieved by combining local knowledge and scientific knowledge towards a common goal (Berkes 2004; Berkström et al. 2019). This kind of knowing together was practiced around our case study in the River Eurajoki, as we found that local ecological knowledge is very spatial and site-specific (see Figure 9.4).

By looking at the living environment of local community residents and collecting their personal as well as shared local ecological knowledge, we open the door to a subjective, experiential landscape. Researchers have little access to this landscape except by meeting people and inviting them to

Figure 9.4 The importance of the Eurajoki lies in its accessibility: the river has been close by, in or behind the home yard, running through the village and bordering the village.

Source: Photograph by Euran kotiseutuarkisto (1960s).

become research partners (Puolamäki 2022, 372–375). In this kind of interaction between researchers and the people they are studying, we need to forget the dichotomy between different kinds of knowledge, such as research knowledge and local ecological knowledge, and focus on what we can know together about the object under study (Höckert 2016):[9]

> We go to the beach every week to see what changes have been made there […] I heard that my father and his friends used to jump off the Kohna Bridge into the water although there were tree snags.
>
> (Eurajoki workshop at Panelia School on 4 September 2019, 12-year-old student)[10]

Local ecological knowledge is often accumulated through everyday experience – it is experiential and first-hand, and collected and processed independently, but also in collaboration with other people. This knowledge is actively reinterpreted, for example, by searching for information in the research literature (Fingerroos and Haanpää 2024). Local ecological knowledge is also assessed and processed at home. In this way, it may be combined across generations. When researchers and the local community share their knowledge, that is, understand and discuss together, research data and local ecological knowledge produce interpretations of the environment that would otherwise be unattainable. Good interaction and dialogue are therefore key to ensuring that both sides have a voice, right up to the point where the research results are made more impressive (See Mäki 2020, 43).[11]

Local ecological knowledge is characterised by its very locality – it engages with knowledge acquired through deep and accumulated everyday experience of, for example, local species, water, soil, and landscape interactions. Interpretations of the environment based on observation and experience that lead to knowledge formation are linked to personal environmental culture and local, social environmental relations. Therefore, local ecological knowledge also has a community dimension (see Figure 9.5); it reflects a local way of living in interaction with the community and the environment (Sedmak 2016; see also Staffans 2004, 67–72). Local ecological knowledge is not absolute or uniform. Different and even contradictory perceptions of the same site, such as the River Eurajoki, can emerge as a result of personal interpretations of the river. However, rather than undermining the reliability or usefulness of local ecological information, these divergent perceptions enrich the picture of the site and its importance to the community.

The importance of local ecological knowledge is not limited to the concrete environmental information and values it provides, such as knowledge of the habitat of species or monitoring data on water quality fluctuations. It enables the identification of intangible environmental values and cultural meanings of the environment, i.e., *cultural ecosystem services*.[12] These are similar in nature to local ecological knowledge, that is, they are relational, contextual, and culturally constructed, arise from multiple cultural sources, and interact

Figure 9.5 Schoolchildren at a workshop.
Source: Photograph by Laura Puolamäki (2019).

with the local environment (Allen et al. 2021; Ahtiainen et al. 2019). Milcu et al. (2013) have noted how mapping the cultural ecosystem services of the environment under study can open up the debate on the reconciliation of different knowledge and values. In this framework, local ecological knowledge has its own relevant role alongside other types of knowledge. It helps understanding opposing views on the shared environment and the root causes of environmental conflicts (Hanaček et al. 2021). Similarly, according to Allen et al. (2021), the interaction and social learning process required to map cultural ecosystem services is in itself as valuable as the resulting outcomes.

However, the challenge of local ecological knowledge is that it is hidden at stages of the decision-making process where solutions are still open. Local ecological knowledge usually only comes to the fore when the options presented are perceived as inadequate or the solutions adopted are wrong. This was also the case with the residents and entrepreneurs of the River Eurajoki. Applications for environmental permits for industrial plants and wastewater treatment plants considered the river as an ecosystem service whose main function is to transport treated wastewater to the sea. The permitting process therefore covers only a part of the ecosystem services, and cultural ecosystem services in particular are excluded from this analysis. Therefore, the limits they impose on the use of the river are not even assessed in the permitting process. When the limits were then exceeded and residents and entrepreneurs stood up to the operators who had applied for the permit, it was a surprise.

The *River Eurajoki Working Group*, which appeared to be an environmental movement, was in fact a cultural movement, which derived its strength from the invalidation of the cultural ecosystem services of the River Eurajoki and the emotional state it created. The application for a permit for a sewage treatment plant in the upper Eurajoki stated that the river had little recreational value. If the cultural ecosystem services of the River Eurajoki had been known at the time of the application for the environmental permit, this assessment might not have been included in the application. However, the application for the environmental permit did cover all the issues required by law. The cultural-environmental relationship has therefore not been identified as a factor that should be taken into account when planning activities targeting the aquatic environment and associated landscape. On the other hand, local ecological knowledge about a single landscape element, such as the River Eurajoki, and the understanding it brings of locally relevant cultural ecosystem services, can only be accessed individually and through encounters organised by someone, either physical or virtual. Reaching that knowledge would therefore have required the process of 'knowing together' mentioned earlier, prior to the initiation of permitting processes.[13]

Discussion

The knowledge held by the residents and entrepreneurs we interviewed in the River Eurajoki basin was primarily of a self-perceived nature, but of course also of a community-derived nature, having been heard and processed. This could include knowledge found, interpreted, and assimilated from archives and research literature, but was above all linked to places that were meaningful to the interviewees themselves and to the local community, and to which they related experiences and stories (Fingerroos and Haanpää 2022, 467; Jokela 2021, 79). In looking at the cultural values of the River Eurajoki, we therefore drew on the local ecological knowledge that had been passed down through generations, and adopted by residents and entrepreneurs. However, this knowledge is of such a nature that it cannot be measured or documented using traditional positivist scientific methods. Rather, it is personal, experiential, and localised, yet it can express values that guide community action, for example (see Hänninen et al. 2005, 3–5, 107; Polanyi 1998, 263–264, 299–304). It is precisely because of this groundedness in experience and practice that knowledge is difficult to process and is perhaps not included in official data on the state of the environment.

In the field studies, we looked at the interactive relationship between the River Eurajoki and the local community. We highlighted the cultural values associated with the water body, both past and present, in order to build an understanding of the historical, social, and cultural significance of the river. This kind of knowledge, produced jointly by researchers and the local community, opens up pathways to the meaning of the landscape surrounding the river, and also builds trust between the different actors who use the river in different ways, as well as *shared ownership* (see Alexander et al. 2016) of the common environment.

Both local ecological knowledge and cultural ecosystem services are thus inherently subjective, contextual, and changeable. However, framing them according to a general and objective measurement framework strips the phenomenon of its fundamental nature, that is, individuality. When discussing local processes in the case of the River Eurajoki, the main significance of one of the processes of knowing is the broad understanding it produces of the landscape element under discussion, namely the river. In this dialogue, different concepts and types of knowledge also merge and complement each other (Puolamäki 2020, 13–18).

In our studies, we also note how a cultural-environmental relationship can trigger individual and community citizen activism, as people feel the need to act on behalf of their environment. Citizen activism, in turn, often only becomes visible in conflict situations, as was the case with the River Eurajoki. Citizen activism, similar to civil disobedience, occurs when people concerned about their environment take action after losing faith in the authorities' ability to care for the natural environment they love (Leino et al. 2010).

Organised civic activity is the voluntary conservation of the environment or nature by individuals, either on their own land or in a perceived shared environment such as lakes and rivers. Walking in one's own environment trains the observation and understanding of nature's own processes and carrying capacity. At the same time, it builds the will and ability to act responsibly. However, the perception of responsible action may be different or even contradictory between different actors, which can lead to a situation where nature's carrying capacity is constantly exceeded.[14] As research on the carrying capacity of nature and the changes taking place in it has increased, the potential of local ecological knowledge has largely remained untapped. We already have the means to highlight different types of environmental knowledge. The next step should be to develop the skills to harness all available knowledge for the benefit of our shared environment.

Notes

1 We have made a conscious choice to use the names of the interviewees in our chapter. Each interviewee was informed orally and in writing about the purpose and objectives of our research, as well as the use, archiving, and possible further use of the data. Consent to the study has been obtained from each interviewee, including permission to use their names. The residents interviewed have chosen to appear openly in the research under their own names: they have felt that their own living environment is valuable and that their own relationship with their environment is worth telling.

2 The majority of the interviewees were born between the 1930s and 1950s, but there are also a few people born in the 1960s and 1970s. They all share an interest in their own environment and the state of the River Eurajoki basin, based on

their profession, lifestyle, and history, or general concerns. Their backgrounds are mainly as home farmers, but also include a teacher and a water engineer.

3 Many interviewees talked about swimming places on the river. See also interviews with Ritva Tuominen 23 April 2014; Marjatta Stenberg and Helena Pinomäki 7 May 2013; Jaakko Heiska 7 May 2013.

4 Liam Campbell's ethnographic research focuses on the River Foyle in Ireland, its history and cultural importance.

5 During the first two years, research data was collected from the River Eurajoki basin through interviews, photographs, and historical maps of the area compared to the contemporary landscape. Two scientific articles (Haanpää et al. 2017, 2018) and one non-fiction book were published from the data, with a multidisciplinary research team including representatives from the humanities and natural sciences. The reference book *Eurajokea silloilta ja siltojen välistä* was published by the Pyhäjärvi Institute in 2022.

6 By citizen science, we mean the participation of research subjects, in our case local community members such as children and adults, as equal researchers and knowledge producers in a field research project. See, for example, Puolamäki 2022.

7 The special group of the Satakunta Centre of Living Image "Meikäläiset" shot a mini-documentary in July 2019. The documentary was screened in Rauma as part of the Blue Sea Film Festival in 2019. This allowed us to involve the special group in our research and to expand our recording methods beyond the limits of our own expertise.

8 Laura Puolamäki has collected her research data from the Eurajoki area between 2011 and 2016.

9 Emily Höckert has applied the concept of hospitality in her study on tourism and ethics, which was carried out in the village of San Ramón, Nicaragua. The concept of hospitality in tourism research is, in our opinion, well-suited to positioning oneself as a co-investigator when working with, for example, children.

10 Seppo Aarikka details how tree logs sometimes formed dangerous debris at bridges crossing the River Eurajoki (interview 10 October 2020). In the same places, sunken and waterlogged logs used to pile up. Swimming stopped in the 1950s. The concept of floating logs is still known along rivers decades after the end of the floating tradition.

11 Maija Mäki used two research methods to build co-researchers among archaeologists and archaeology enthusiasts working in the field of cultural tourism. According to Mäki, the researcher is forced to reposition themselves in a partnership of scholarship as the work progresses. The outcome of the interviews is shaped by the collaboration between the researcher and the interviewee, and influenced by the interaction that is built up during the research process. Practices of hospitality are present in this interaction.

12 The Millennium Ecosystem Assessment (2005) assessed the importance of nature for human well-being by describing it as an ecosystem service. Cultural ecosystem services include intangible services provided by the natural environment that are related to the cultural and social relationships with the environment. However, the extensive report mentions cultural ecosystem services only a few times.

13 Community workshops have proven to be the best method of highlighting locally significant values when exploring the cultural ecosystem services associated with mountain lakes.

14 Earth Overshoot Day has moved from December to July in the past 50 years. Earth Overshoot Day marks the date when humanity has exhausted nature's budget for the year. For the rest of the year, we are maintaining our ecological deficit by drawing down local resource stocks and accumulating carbon dioxide in the atmosphere. We are operating in overshoot (Global Footprint Network 2026).

Bibliography

Aarikka, Seppo, interviewed by Laura Puolamäki, 10 October 2020. Degree Programme in Digital Culture, Landscape and Cultural Heritage.

Abrams, Lynn. 2010. *Oral History Theory*. London: Routledge.

Ahtiainen, Heini, Eero Liski, Eija Pouta et al. 2019. "Cultural Ecosystem Services Provided by the Baltic Sea Marine Environment." *Ambio* 48: 1350–1361. https://doi.org/10.1007/s13280-019-01239-1

Allen, Karen E., Celia Castellano and Sophia Pessagno. 2021. "Using Dialogue to Contextualize Culture, Ecosystem Services, and Cultural Ecosystem Services." *Ecology and Society* 26 (2): 7. https://doi.org/10.5751/ES-12187-260207

Alexander, Steven M., Mark Andrachuk and Derek Armitage. 2016. "Navigating Governance Networks for Community-based Conservation." *Frontiers in Ecology and Environment* 14 (3): 155–164. https://doi.org/10.1002/fee.1251

Anttila, Lauri, Riina Haanpää, Eeva Karhunen, Teija Kirkkala, Laura Puolamäki, Eeva Raike and Henna Ryömä. 2022. *Eurajokea silloilta ja siltojen välistä*. Eura: Pyhäjärvi-instituutti.

Aswani, Shankar, Anne Lemahieu and Warwick H.H. Sauer. 2018. "Global Trends of Local Ecological Knowledge and Future Implications." *PLoS ONE* 13 (4): 1–19. https://doi.org/10.1371/journal.pone.0195440

Aziz, Gabriele Volpato, Michele Filippo Fontefrancesco and Andrea Pieroni. 2022. "Perceptions and Revitalization of Local Ecological Knowledge in Four Schools in Yasin Valley, North Pakistan." *Mountain Research and Development* 42 (4): R1–R9. https://doi.org/10.1659/MRD-JOURNAL-D-21-00061.1

Bagayoko, Sidylamine and Sanna Tawah. 2014. "Vallankäyttöä linssin edessä ja takana: visuaalinen etnografia ja kaksi tutkimusta Länsi-Afrikasta." In *Moniulotteinen etnografia*, edited by Pilvi Hämeenaho and Eerika Koskinen-Koivisto, 185–211. Helsinki: Ethnos ry.

Berkström, Charlotte, Myron Papadopoulos, Saleh Jiddawi Narriman and Lina Mtwana Nordlund. 2019. "Fishers' Local Ecological Knowledge (LEK) on Connectivity and Seascape Management." *Marine Fisheries, Aquaculture and Living Resources* 6 (2019). https://doi.org/10.3389/fmars.2019.00130

Berkes, Fikret. 2004. "Rethinking Community-based Conservation." *Conservation Biology* 18 (3), 621–630.

Bläuer, Auli. 2019. "Eurajoen keskiaikaisen Liinmaan linnan luuaineisto – yhteenveto." In *Puukenkien kopinaa, Henrik Asplundin juhlakirja*, edited by Janne Harjula, Visa Immonen and Juha Ruohonen, 261–275. Turku: University of Turku.

Campbell, Liam. 2021. *Room for the River. The Foyle River Catchment Landscape: Connecting People, Place and Nature*. Derry: Merdog.

Charnley, Susan, A. Paige Fischer and Eric T. Jones. 2007. "Integrating Traditional and Local Ecological Knowledge into Forest Biodiversity Conservation in the Pacific Northwest." *Forest Ecology and Management* 246 (1): 14–28. https://doi.org/10.1016/j.foreco.2007.03.047

Eurajoki-workshop, Panelia School, Eura, interviewers Laura Puolamäki, Eeva Karhunen and fifth and sixth grade pupils, 4 September 2019. Degree Programme in Digital Culture, Landscape and Cultural Heritage.

Eurajoki-workshop, Irjanne School, Eurajoki, interviewers Laura Puolamäki, Eeva Karhunen and fifth and sixth grade pupils. 23 September 2019. Degree Programme in Digital Culture, Landscape and Cultural Heritage.

Falttu, Heikki, interviewed by Anna Kimpimäki and Taru Äkräs, 23 April 2014. Degree Programme in Digital Culture, Landscape and Cultural Heritage.

Fingerroos, Outi and Riina Haanpää. 2022. "Yhteisölähtöiset muistitietohankkeet osana yliopisto-opetusta." In *Muistitietotutkimuksen kentät: Teoriat, käytännöt ja muutos*, edited by Ulla Savolainen and Riikka Taavetti, 459–480. Helsinki: Finnish Literature Society.

Fingerroos, Outi and Riina Haanpää. 2024. "Yhteisöllinen muisti ja paikallinen tieto." In *Osallistuminen ja osallisuus kulttuurisuunnittelussa: menetelmät, teoria ja käytäntö*, edited by Kaisu Kumpulainen and Eerika Koskinen-Koivisto, 132–147. Sophi: University of Jyväskylä.

Haanpää, Riina. 2014. "Haastattelumenetelmä ihmisten ympäristökokemusten kartoittamisessa." In *Osana omaa ympäristöä,* edited by Laura Puolamäki and Eeva Raike, 24–45. Turku: University of Turku.

Haanpää, Riina, Laura Puolamäki and Eeva Raike. 2017. "Eurajoen muuttuvat arvot." In *Yhteiskuntaetnologia*, edited by Outi Fingerroos, Sanna Lillbroända-Annala, Maija Lundgren and Niina Koskihaara, 240–264. Helsinki: Finnish Literature Society.

Haanpää, Riina, Laura Puolamäki and Eeva Raike. 2018. "Tapaus Eurajoki – ihmisen ja kulttuuriympäristön vuorovaikutussuhde tutkimuskohteena." In *Soveltava kulttuurintutkimus*, edited by Pilvi Hämeenaho, Tiina Suopajärvi and Johanna Ylipulli, 180–201. Helsinki: Finnish Literature Society.

Haanpää, Riina, Laura Puolamäki and Eeva Raike. 2019. "Ympäristön lukutaidon kehittyminen kulttuuripolkujen avulla." In *Maailmanperintö ja kulttuurikasvatus – World Heritage and Cultural Education*, edited by Mika Metsärinne, Tapio Heino, Riitta Korhonen and Maija Esko, 134–147. Rauma: Rauman opettajankoulutuslaitos.

Hämeenaho, Pilvi and Eerika Koskinen-Koivisto. 2014. "Etnografian ulottuvuudet ja mahdollisuudet." In *Moniulotteinen etnografia*, edited by Pilvi Hämeenaho and Eerika Koskinen-Koivisto, 7–31. Helsinki: Ethnos ry.

Hanaček, Ksenija, Johannes Langemeyer, Tatyana Bileva and Beatriz Rodríguez-Labajos. 2021. "Understanding Environmental Conflicts through Cultural Ecosystem Services – The Case of Agroecosystems in Bulgaria." *Ecological Economies* 179: 106834. https://doi.org/10.1016/j.ecolecon.2020.106834

Hänninen, Sakari, Jouko Karjalainen and Tuukka Lahti. 2005. "Johdanto." In *Toinen tieto. Kirjoituksia huono-osaisuuden tunnistamisesta*, edited by Sakari Hänninen, Jouko Karjalainen and Tuukka Lahti, 3–9. Helsinki: Stakes.

Heikkilä, Urpo. 2000. *Vuosisata elämää Eurajoen varressa.* Eura: Omakustanne.

Heiska, Jaakko, interviewed by Kaisa Lindewall and Heli Mäki-Aho, 7 May 2013. Degree Programme in Digital Culture, Landscape and Cultural Heritage.

Höckert, Emily. 2016. "Ethics of Hospitality – Participatory Tourism Encounters in the Northern Highlands of Nicaragua." *Matkailututkimus* 12 (1): 79–84.

Jokela, Timo. 2021. "Kuerin matkassa: paikkaperustainen taideteos ympäristökonfliktissa." *Research in Arts and Education* 2021 (3): 73–97. https://doi.org/10.54916/rae.119563

Jokihelmisimpukka/Margaritifera margaritifera. https://wwf.fi/elainlajit/jokihelmi simpukka/

Kortelainen, Kaisu. 2008. *Penttilän sahayhteisö ja työläisyys. Muistitietotutkimus.* Helsinki: Finnish Literature Society.

Lähteenmäki, Aino and Kaino Nurminen, interviewed by Johanna Roslund and Heli Lehti, 23 April 2014. Degree Programme in Digital Culture, Landscape and Cultural Heritage.

Leino, Helena, Juha Peltomaa and Minna Santaoja. 2010. "For the Sake of the Lake: Multi-Signification and Improvisation in the Practices of Two Lake Restoration Projects in Finland." *Local Environment,* 15 (9–10): 775–784. https://doi.org/10.1080/13549839.2010.509385

Mäki, Maija. 2020. *Polkuja esihistoriasta tulevaisuuksiin. Matkailun tulevaisuuksia ja toimijoita arkeologisilla maastokohteilla.* Helsinki: Suomen muinaismuistoyhdistys.

Merleau-Ponty, Maurice. (1945) 1999. *Phénoménologie de la perception.* Paris: Gallimard.

Merleau-Ponty, Maurice. 2002. *Phenomenology of Perception.* Translated by Colin Smith. London: Routledge.

Milcu, Andra Iona, Jan Hanspach, David Abson and Joern Fischer. 2013. "Cultural Ecosystem Services: A Literature Review and Prospects for Future Research." *Ecology and Society* 18 (3): 44. https://doi.org/10.5751/ES-05790-180344

Pinomäki, Helena and Marjatta Stenberg, interviewed by Leena Kouhia and Eija Westerbacka, 7 May 2013. Degree Programme in Digital Culture, Landscape and Cultural Heritage.

Polanyi, Michael. 1998. *Personal Knowledge. Towards a Post-Critical Philosophy.* London: Routledge.

Puolamäki, Laura. 2017. "Tracing Cultural Landscape Values of Children with Geographic Information System." *European Countryside* 9 (2): 375–396.

Puolamäki, Laura. 2020. *Kätketty maisema. Arkitieto lähiympäristön kohtaamisessa.* Turku: University of Turku.

Puolamäki, Laura. 2022. "Eurajokea etsimässä: Jaettu ympäristösuhde ja lapset tutkijakumppaneina." In *Kulttuurihistorian tutkimus. Lähteistä menetelmiin ja tulkintaan*, edited by Rami Mähkä, Marika Ahonen, Niko Heikkilä, Sakari Ollitervo and Marika Räsänen, 371–386. Turku: Kulttuurihistorian seura.

Roth, Klaus. 1997. "Rivers as Bridges – Rivers as Boundaries. Some Reflections on Intercultural Exchange on the Danube." *Ethnologia Balkanica* 1: 20–28.

Sedmak, Clemens. 2016. "Onions, Soils, and Watering Plants: Local Knowledge and Local Epistemic Justice." In *Appreciating Local Knowledge*, edited by Elisabeth Kapferer, Andreas Koch and Clemens Sedmak, 5–20. England: Cambridge Scholars.

Shopes, Linda. 2002. "Oral History and the Study of Communities. Problems, Paradoxes, and Possibilities." *The Journal of American History* 89 (2): 588–598.

Staffans, Aija. 2004. *Vaikuttavat asukkaat: vuorovaikutus ja paikallinen tieto kaupunkisuunnittelun haasteina.* Espoo: Teknillinen korkeakoulu, Yhdyskuntasuunnittelun tutkimus- ja koulutuskeskus.

Suomi, Tapani, interviewed by Vilho Pirttijärvi and Eeva Louhela, 7 May 2013. Degree Programme in Digital Culture, Landscape and Cultural Heritage.

Tolvi, Mikko and Timo Tolvi, interviewed by Roosa Halonen and Johanna Lumme-Tamminen, 6 May 2013. Degree Programme in Digital Culture, Landscape and Cultural Heritage.

Ukkonen, Taina. 2000. *Menneisyyden tulkinta kertomalla. Muistelupuhe oman historian ja kokemuskertomusten tuottamisprosessina.* Helsinki: Finnish Literature Society.

Uotila, Kari, Hannele Lehtonen and Janne Haarala. 2011. *Eurajoki Liinmaan linna. Arkeologinen tutkimus 1.9.–7.10.2005.* SuVi-projekti. Turku: University of Turku.

Valla, Jenni. 2019. *Vanha Turku oli kalju, haiseva ja likainen*. https://www.auroralehti.fi/vanha-turku-oli-kalju-haiseva-ja-likainen/

Virtanen, Katrina. 2022. "Luonnon ja ympäristön taju." In *Virtaava sydän, Euran ja euralaisten historiaa 1860-luvulta 2000-luvulle*, edited by Anna Sivula and Anni Ruohomäki, 41–47. Eura: The Municipality of Eura.

Yli-Pelkonen, Vesa and Johanna Kohl. 2005. "The Role of Local Ecological Knowledge in Sustainable Urban Planning: Perspectives from Finland." *Sustainability: Science, Practice and Policy* 1(1): 3–14. https://doi.org/10.1080/15487733.2005.11907960

10 Sensing Place(s) in the Humilocene

Explorations in Ecological Storytelling

Ullrich Kockel

Approaches to including non-human worlds in analyses of the current polycrisis have been developed under various labels, such as 'more-than-human', 'other-than-human', 'anti-human', 'posthuman', 'transhuman', 'multispecies', or 'transspecies'. These conceptual frameworks are usually understood as perspectives addressing the devastation caused by the Anthropocene. While they 'have lacked clarity in their definitions, interpretations, purposes, uses, and range of application', they are offered as concepts that 'guide us to transform our human-self-consciousness around a pluralist ethos that welcomes others' ideals, voices, perspectives, and modes of experiencing the world' (Ducros 2021). A common denominator is the challenge to anthropocentric perspectives. De-centring *Anthropos* from our perspective on *Oikos* is not only commendable but also acutely necessary as we are beginning to understand the damage a particular type of anthropocentric worldview has wrought, especially over the past couple of centuries, to the world it has been viewing. With growing recognition that even the Anthropocene belongs with this category of worldview (see Chapter 1 in this book), there have been calls for a better conceptual framework. David Abram's (2020) postulate of a *Humilocene* marks a significant step away from the anthropocentric misconception that the Anthropocene is. I will come back to this towards the end of these explorations.

Towards a Creative Ethnology

As a European ethnologist and empirical cultural scientist with some historical awareness, I find in this 'posthuman' turn of discourse resonances of the emic/etic debate of some two generations ago, and wonder: to what extent can we actually leave behind the existential condition of our inevitably *emic* perspective and slip into an *etic* one that observes our world quasi from outside ourselves? Is the encouragement of looking at the world through 'other-than-human' eyes not merely an invitation to anthropomorphise that world in new ways? As topographers conducting our participant observation in multispecies places, can we truly expect to understand, and faithfully convey, say, 'squirrelness', or an oak's sense of belonging in a deciduous grove;

DOI: 10.4324/9781003727972-10

and how, other than poetically, can we know that a river is really 'carried by the surprise of its own unfolding' (O'Donohue 2000, 30)? Or are we just having a good time pretending? And, as the authors ask in Chapter 1 of this book, would that really be such a bad thing to do?

At the time of writing, I am exploring the potential of ecological storytelling – a genre of what in Scotland is called 'creative ethnology' (Kockel and McFadyen 2019) – for critical cultural analysis. My perspective is informed by historical ways of studying ecological lifeworlds – further inspired by eco- and geopoetic aspirations to shape these lifeworlds into 'home' (however defined) – and drawing on some three decades of personal research into issues of place, landscape, and belonging in a variety of human-ecological contexts across Europe, from metropolitan areas to contested borderlands, rural, and maritime regions. In this chapter, I am sharing some of these explorations. Following an outline of theory and method in creative ethnology, I consider aspects of emerging multispecies perspectives before turning to an example of how these approaches can be applied creatively to stimulate sense of place in an urban context: the 'Two Rooms' exhibition at the Vilnius City Museum.

Museums have long been a playground for ethnological practice, displaying selections from their collections of objects according to contemporary theoretical paradigms and educational objectives (Kockel 2015a; Kockel and Nic Craith 2015). As 'complex places of learning, exchange and experience [...] the roles of museums are being rethought' (Fenske and Elpers 2019, 8). Contemporary museums are seen not so much as display windows showcasing their collections, but increasingly in terms of capturing and narrating the varied ecological contexts of their exhibits (Nic Craith and Kockel 2016), becoming 'sites for rearticulation and change' (Fenske and Elpers 2019, 8), and indeed 'agents of transformation' (Kirshenblatt-Gimblett 2019). There is growing emphasis on the active engagement of the public as 'cultural participants [...] co-creating exhibits and collections' (Fenske and Elpers 2019, 8). Curators are challenged to find ways of interpreting and translating multispecies perspectives that take account of the plurality of relationships encompassed by these perspectives.

Rambling Ethnologically – Walking as Method

Many of the early ethnologists who, during the eighteenth and nineteenth centuries, were focussing on domestic fields rather than on exotic – usually colonial – locations studied the lifeworld around them by what Germans called *erwandern* – rambling and roaming (e.g., Riehl 1857). Later, in the twentieth century, this became *erfahren* – not so much literally, in the sense of travelling by modern means of transport (although that became increasingly, and necessarily, part of it), but as the gathering of *Erfahrung*, experience. The etymological link indicates an epistemological one – experience is gained through moving, travelling around in a manner that implicitly engages with the world as we go. These approaches to gaining knowledge of the world around us and understanding its constitution, less from reading books and more from direct intellectual and, importantly, emotional engagement, anticipated – and indeed have many parallels with – recent approaches to developing ecologically grounded perspectives that emphasise sense of place, belonging, and interconnectedness, such as Human Ecology (e.g., Williams et al. 2012), Environmental Humanities (e.g., Rose et al. 2012), Green Storytelling (e.g., Molthan-Hill et al. 2020), or Geopoetics (e.g., Malpas and White 2021; Kockel 2023). However, Lesley Instone (2015, 134) noted the extent to which her 'practices of movement and thinking are all shaped by historically contingent cultural practices of looking and moving that are familiar to those with a European cultural heritage'. This suggests that cultural experience and cultural expressions remain fundamental aspects of human life in all of this – which pragmatically presupposes the existence of 'culture' (de Munck 2004), despite many critics of that concept (e.g., Kuper 1999). We cannot simply get away from the actuality that our perception is mediated, culturally conditioned.

Over the past few decades, academic interest has gradually moved from 'sense of place' more generally, first problematised by geographers like Yi Fu Tuan (1974) and philosophers like Edward Casey (1996), towards the constituent elements of 'home' in different cultural contexts (e.g., Bahun and Petrić 2018). My own cultural context is originally German, and my focus has, not surprisingly therefore, been on the peculiarly German concept of *Heimat* – on whether, how, and where this can be found, founded, and created, and what an ethnology that draws on the creative arts, including the traditions it studies, may contribute to that process.

There are deep, multi-layered historical connections between Creative Ethnology and a Human Ecology defined as 'a transdisciplinary science of human populations and their cultures, treating them as a creative element of ecosystems and social systems [...] attracting attention to the role of environment in the formation of human biological and cultural characteristics' (Wolanski and Henneberg 2001, 3). Ever since being a doctoral student in social and environmental studies, I have argued for a re-orientation and renewal of ethnology, proceeding from deep ecological foundations that might – if somewhat controversially – be regarded as an indigenous approach to studying local culture – culture in, of and with place. 'Ethnology' can be translated into my native German as *Volkskunde* – 'knowledge of (the) people', in both its singular and plural meanings. The *International Encyclopedia of the Social Sciences* defines 'ethnology' as a discipline that theorises 'about the persistence, adaptation, and function of tradition within complex societies', offering 'concepts for the way that individuals and groups use tradition to express values, beliefs and ideas in a number of forms, including art, architecture, story, song, speech and custom'. Historically, 'ethnology has a legacy of anthropological attention to the social and material basis of tradition, particularly in rural and peasant societies (especially in Scandinavia and Germany where holistic terms *folkliv* and *Volkskunde*, respectively, circulated)'

(Bronner 2023; orig. italics). Understood in this way, 'ethnology' can be seen as an approach to *the local*, promoting a comparative understanding of the *own* by, through and, crucially, with the *other* – meaning both humans and non-human subjects, viewed as part of a *local household* (*oikomene*). It thus provides grounding for what, with another controversial German term, I describe as *Heimatkunde*, the knowledge of *Heimat* – the place that we may possibly be from, but are always towards (see Bloch 1974; also Greverus 1972). Postulated originally around the same time as a scientific-technical human ecology, *Heimatkunde* can be regarded as its spiritual complement *sine qua non*, facilitating an ethnologically informed human ecology of place and belonging (Kockel 2012a, 57), providing radical foundations for telling new stories of 'home' that facilitate the conversion of world into *Heimat* (Bloch 1974) – a relational way of dwelling that overcomes our alienation and 'lack of earthly self-awareness' (Milstein and Castro-Sotomayor 2020, xvii). This act requires turning the focus of ethnology from its traditional 'salvage' approach (see Nic Craith 2020) aiming to archive and musealise, towards what a forward-looking tradition can create by drawing on past knowledge, experience, and stories – that is, towards a creative ethnology (Kockel and McFadyen 2019) aware of its responsibilities.

Telling a Story

We are living in a time of polycrisis (Fenske 2023, 2f.) – climate change; loss of biodiversity; soil degradation; pollution; poverty; famine; steady growth of the human population; growth of social inequality; socio-economic separation; populism – to name but a few of the multiple crises we are facing. Moreover, the many different aspects of multiple crises are deeply intertwined: you cannot solve one crisis without considering another. The loss of biodiversity is closely connected to growth in the number of humans and increasing social inequality. Scholars have suggested different concepts to analyse these multiple crises, the most commonly used being the 'Anthropocene', which emphasises human influence. It 'not only marks a geologic time, but also a scholarly one – a moment when a particular term has captivated scholarly imaginations' (Swanson et al. 2015, 150).

Donna Haraway, in her *Companion Species Manifesto* (2003), spoke of 'naturecultures' as an emergent perspective and evoked a new multispecies anthropology, the study of humans and non-humans in their interrelatedness. To adequately address environmental problems, climate change, and biodiversity issues, but also concerns of humanity's health and nutrition, the social sciences, and humanities cannot limit their scope to looking exclusively at human beings, their cultural meanings, and social practices. Anthropology, once regarded as 'the human science', is on its way to becoming a more-than-human concern (Welz 2021, 38).

Whatever concept one prefers to interpret our times, two aspects appear to be common to all of them (Fenske 2023, 3) – the tremendous importance of narratives, and the awareness that stories with a happy ending have become rare nowadays. Donna Haraway invites us to make kin with, and rediscover, our non-human relatives. She also argues that it matters what stories we tell. Moreover, according to Haraway and many other scholars, our multispecies societies need new stories – stories of loss that mourn previously existing chances and possibilities; stories of resilience that reflect new perspectives; stories of hope that enforce our confidence in future solutions to the many dilemmas we currently face (see Fenske 2023).

Narratives are media of pivotal importance to understanding the world, to learn how people perceive changes, how they shape changes, how they react to changes. Narratives allow us to understand who we are, how we have become what we are, and how we may shape our future world. Narratives, therefore, are powerful tools to subvert the status quo and enable change.

The scholarly study of narratives has a tradition of more than two hundred years, connected closely with the early ethnologists' practice of *erwandern*. Nowadays, the study of narratives is revived and invigorated by such new fields as the ethnography of narrating, or research on narrative culture (e.g., Nic Craith 2020). Researchers in ethnology understand narratives and narrating as a mode of comprehension, applied as an instrument for change. In August 2023 ethnologists at the University of Würzburg in Germany hosted an international conference on 'Narrating the Multispecies World', to share and discuss some key questions (Fenske 2023, 9):

- What are the stories of our multispecies world today? Are there stories of our multispecies world? Are people aware of other than humans and do they narrate about them? Do people narrate climate change?
- Who is narrating what stories for whom, where, when?
- Which observations, needs, desires, dreams, nightmares, aspirations, and ethics are shared by narrating?
- What is the role of the past? Which parts of our narrative heritage do we still maintain?
- What are the new powerful stories developing possibilities for a peaceful cohabitation in the multispecies world?

The contributors to the *Manifesto for Living in the Anthropocene* (Gibson et al. 2015) argued that stories enact connectivity, entangling us in the lives of others; have the capacity to reach beyond abstractions and move us to concern and action; are rich sources of reflection; and enliven moral imagination, drawing us into deeper understanding of responsibilities, reparative possibilities, and alternative futures. Narratives are important repositories of values, political and ethical ideas, and behaviours that determine how we engage with our ecological homes. They suggest that changing the way we interact with the environment requires not only new stories but also a better understanding of the ones that have long been in circulation.

Stories produce bodies that produce stories in an endless intra-active metabolic continuum. They not only represent material worlds but also shape and make new worlds. From these premises, Hohti and Tammi (2023) develop what they call 'composting storytelling,' an approach to heterogeneous, open-ended, small stories, interwoven with everyday interaction, that draw attention to the polyphony of voices and temporalities, and put the human as individual agent in critical perspective.

Rose and van Dooren (2012, 1) noted that '[o]ver fifty per cent of the world's human population now lives in cities, and the rate is increasing exponentially'. We do not need to be overly precise in our definitions here; I recall a speaker at a geography conference in the late 1980s claiming a similar demographic figure; from my own fieldwork I knew that the Irish Census at the time defined 'urban environment' as settlements of more than 1,500 people. But population size is not the point here – the proliferation of built-up areas and tarmacked surfaces is.

To draw attention to animals in urban spaces is nothing new. At the same time humans are migrating to cities, or finding their rural homes overtaken by urban development, many other animals are also making the transition to urban life. For them, too, reasons for urbanisation, while complex, include the same two major constellations of causes: animals are choosing to move into urban spaces, and they are finding their rural places overtaken by growing urban spaces. Meanwhile, increasing interest in 'biodiversity' and 'urban nature' has made humans aware like never before of the many other animals that reside in our urban environments – some newly arrived, others pre-dating the urban development itself. Rose and van Dooren's thinking here is rooted in a notion of place as relationally constituted, resembling the German *Heimat* as considered earlier, or the Gaelic *dúchas* (Nic Craith, in this volume). Places are materialised as historical and meaningful, and no place is produced by a singular vision of how it is or might be. Places are co-constituted in processes of overlapping and entangled 'storying' in which different participants may have very different ideas about where we have come from and where we are going (Rose and van Dooren 2012, 2).

Since the early twentieth century, Estonian biologist Jakob von Uexküll's concept of an organism's *Umwelt* – the recursive perceptual lifeworld characterising an organism and its surrounds – has profoundly influenced several generations of thinkers. Rose and van Dooren (2012, 3) drew on this growing body of research, directing it toward a particular focus on story and its intra-actions with place. They were working with a broad notion of 'story', meaning that which emerges out of an ability to engage with happenings in the world as sequential and meaningful events. Several chapters in the present book consider such creative engagement, whether by way of conversation or material interaction.

An analysis that links story, place, and the non-human world confronts fascinating questions: how would we (humans) know an animal is entangled in a storied place and is participating in, shaping, and being shaped by, that story? Rose and van Dooren (2012, 5) looked at narrative as action – where animals go, and what they do – proposing that in many instances an animal's actions articulate a narrative of place, indicating the inhabitation of a storied world, and thus embedding the animal in place just as we humans do with our narratives of belonging.

Embedded in Place

For many years, my answer to the question: 'Where are you from?' used to be along the lines of 'nowhere in particular'. Depending on their mood and general attitude to life, people interpreted this answer somewhere along a continuum ranging from 'intriguingly cryptic' to 'annoyingly evasive'. Consequently, it could spark off an animated discussion of place and belonging, or stop a pleasant conversation dead in its tracks, the difference often depending on the enquirer's autobiographical experience of place. For me, however, the response was neither a matter of evading the question nor an attempt to shroud my life in mystery, but a rational approach to the dislocation that has been so widely hailed as the essential existential condition of the postmodern human being – usually by voices that in the same breath decry the very notion of essential anything (Kockel 2012a, 58).

As an anthropologist, I owe much to the late Ina-Maria Greverus, founding professor of the Frankfurt Institute of Cultural Anthropology. Her often-contested interpretation of 'human ecology' (Greverus 1972) went on to influence later generations of researchers in our field whose pre-occupation was – and in some cases, like mine, continues to be – with human-ecological relations. In her writings on cultural ecology in 1970s Germany, Ina-Maria Greverus asserted that the term 'environment' comprised both the natural and the human-made, and that culture needed to be considered both the product and the process of humans who are collectively interacting with their environment. Greverus was among the first scholars in the humanities in post-war Germany who dared to link ethnological research in modern societies with key issues of biophysical anthropology. Her pioneering work on patterns of human territoriality, however, did not build on earlier, deeply problematic German intellectual traditions, but took inspiration from cultural ecology, a fairly new subfield of anthropology that had emerged in the United States, where cultural anthropologists had already in the mid-twentieth century started developing new perspectives on humans as a biological species, and on human–environment relations. Drawing on early human ecology, they tried to bridge the ontological divide between anthropology as a generalising science of human existence and the detailed ethnographic studies of local lifeworlds that increasingly had become the hallmark of their discipline. For Greverus, this emerging anthropological subfield, variously called human ecology or cultural ecology, became an important influence. Cultural ecologists regard culture as the specifically human way of adapting to environmental conditions, enabling them to survive. Their perspective closely reflects that of the polymath Patrick Geddes, ecologist and 'father of town planning', whose intellectual contribution to the Scottish Renaissance at the turn of the twentieth century has been described as having been at once modernist and revivalist (White et al. 2023, 4). For Geddes, 'a sustainable future required an understanding of the past [...] and his modernism did not simply learn from the past, it depended on it' (Macdonald 2020, 146).

Geddesian thinking underpinned my postulate of an approach to understanding the composite of place ethnologically via a cyclical triad (Kockel 2009) of Topography (careful description of place; detailed account of 'here') from which a Topology (knowledge of place; place-based policy and practice) can be built up to nurture an intercultural Toposophy (wisdom of place; philosophical understanding) that in turn informs Topography (see Figure 10.1).

The term Toposophy, literally 'Place Wisdom', was derived from Keith Basso's (1996) wonderful study among the Western Apache, 'Wisdom Sits in Places', and inspired the foundation of the Place Wisdom working group in the International Society for Ethnology and Folklore (Kockel 2010). Writing at about the same time as Marc Augé (1995) proposed his ethnology of solitude in response to the non-places he diagnosed as the settings of our 'supermodern' condition characterised by growing displacement and up-rooting, Basso (1996, 109) argued that:

> it is simply not the case [...] that relationships to place are lived exclusively or predominantly in contemplative moments of social isolation. On the contrary, relationships to places are lived most often in the company of other people, and it is on these communal occasions – when places are sensed together – that native views of the physical world become accessible to strangers.

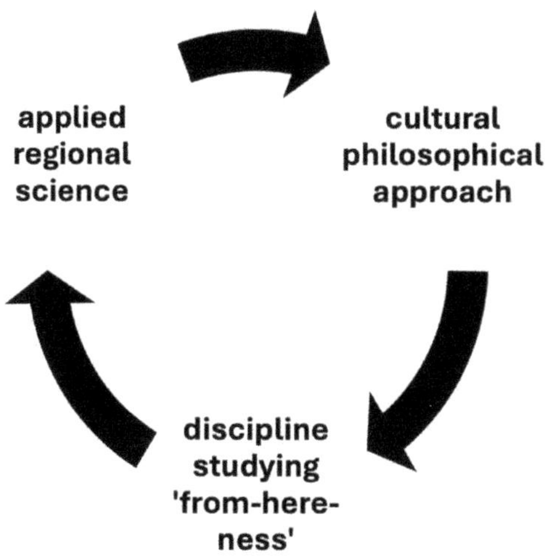

Figure 10.1 Ethnologically understanding the composite of place.

Source: Figure by Ullrich Kockel (2009).

Discussing the increasing cultural displacement of Native American youth as they lose their local embedding through intergenerational stories they no longer share, Basso (1996, 145) observed that the crucial question concerning sense of place is 'not where it comes from, or even how it gets formed, but what [...] it is made with'. In other words, its character 'emerges from the quality of its ingredients'. Place is not just spatial but, importantly, deeply temporal (Kockel 2012b, 565).

The twentieth century has been described as 'the century of displacement' (Dakhli, Laborier, and Wolff 2024). This label refers especially to the forced migration of humans from their homelands in the course of local or global conflicts, and my own research among expellees and refugees from former German cultural territories after the Second World War and their descendants (Kockel 2015b, 2019b) has been in this context. However, their life stories have also raised other, related displacements: that of animals and plants becoming extinct; landscapes – both rural and urban – changed beyond recognition; and individuals becoming disconnected (also by factors other than violent armed conflict) from stories that rooted them, and unable to grow fresh roots in their location – which may not even have changed much physically – due to a lack of credible new stories. These displacements have been experienced by other people in other lifeworld contexts, such as migrants despairing over the steady receding of their transnational utopia, or communities on the Atlantic periphery whose bioregional ecology – including the physical environment, traditional livelihood, and even language – is gradually being eroded by economic and climate change.

Common Grounding

Multispecies Studies understands organisms as situated within deep, entangled histories. Drawing inspiration from the natural sciences and beyond, it brings these diverse bodies of knowledge into conversation and pushes them in new directions. Multispecies scholars are asking how human lives, lifeways, and accountabilities are folded into these entanglements. In taking up these questions, scholars are also engaging with long histories of relational, agentic thinking from indigenous peoples (van Dooren et al. 2016, 2). The latter has become one of my major research topics, with particular focus on the traces of traditional ecological knowledge, broadly conceived, surviving in European contexts (e.g., MacDhonnchaidh and Robertson 2022), and on the potential re-rooting of displaced people through proactive entanglement in place (Kockel 2012a, 2012b), aided by interventions of a creative ethnology.

The late Kenneth White's *geopoetics* (Malpas and White 2021) has been an important inspiration in my quest for such a creative ethnology (Kockel 2023). White has also drawn attention to the 'social sculpture' concept of German artist Joseph Beuys as a form of pursuing the proactive entanglement in place (Kockel 2011). Of course, 'artists have long generated multimedia installations and performative interventions to bring attention to animals, plants, fungi, and others at the periphery of anthropocentric worlds' (van Dooren et al. 2016, 9).

This represents 'the arts of attentiveness', involving 'both a practice of getting to know another in their intimate particularity [...] and, at the same time, a practice of learning how one might better respond to another, might work to cultivate worlds of mutual flourishing' (van Dooren et al. 2016, 17). Several contributions to the present book highlight such practice, which is not entirely without precedents (Kirksey and Helmreich 2020), many of which, as the early ethnologists, predate the disintegration of academic research into fields and disciplines, and the growing detachment of these from actual everyday entanglements in the interest of a 'rigorous inquiry' seeking universal truths.

As anthropologists, we tend to recognise cultural ascriptions as relative and specific, but their incidence as universal. The process is social in the sense of societal location, but we are nevertheless dealing with a cultural system of meaning that can be regarded as place-specific, and indeed often makes sense only if understood in this specificity (Kockel 2012b, 561). Memories establish and maintain place relationships; they 'replace' us (Clingerman 2011, 151). Place emerges 'out of activity that does not have the "production of place" as such as its avowed goal' (Weiner 2001, 15), but this does not mean that it is a random creation. Rather, place and its community can be regarded as work of art, as Weiner indicates, pointing to Martin Heidegger's 'concern with the role of [...] aesthetic, form-inducing processes in human social and cultural life in general and not just with respect to [...] artwork narrowly conceived'. Place and community thus are, in a Beuysian sense, 'social sculpture' (see Walters 2012), and 'home' is not so much where we are from, but a future place we are coming to by making it for the first time (see Bloch 1974).

'Remembering place rightly [...] means more than having nostalgia for the pristine wilderness of some mythic past' (Clingerman 2011, 154). The 'complex affinities' of place are a matter of community engagement much more than geography; Basso (1996, 148) argued that the 'social and moral force' of a sense of place 'may reach sacramental proportions, especially when fused with prominent elements of personal and ethnic identity'. Remembering rightly is an ethical challenge, therefore, particularly for groups that may have to come to terms with a past of nationalism and colonialism (Kockel 2012b, 566). This, more than anything, stands in the way of adapting indigenous cosmologies, particularly in a European context, where a rampant populism feeds on visions of a glorious national, colonial, and often imperial past to be recovered (Kockel 2024), a past that was based on the wholesale displacement of 'others' perceived as not fitting the narrative.

Belonging in a Multispecies City: Telling the Story of Vilnius

Arguably, biodiversity raises primarily cultural issues, 'questions of what we value and what stories we tell, and only secondarily issues of science' (Fenske 2023, 9).

The way we tell our own stories is changing, as a brief look at the Vilnius City Museum illustrates. With the working title 'Two Rooms', curators have developed two provisional galleries as part of a planned permanent exhibition at the museum's anticipated new location. This future exhibition will seek to represent five key themes: (1) the city's changing political role within Lithuania over time; (2) the multi-cultural demography of Vilnius; (3) urban planning and architecture; (4) the city's geological setting; and (5) Vilnius residents. It is the latter two that the provisional galleries are showing.

The themes of galleries 1–3 are not new, and have attracted a significant amount of scholarship, from broad-ranging mainstream field study projects to innovative explorations of specific aspects, to comprehensive cultural-geographical histories of the city. Schulze et al. (2010) investigated how Vilnius as a focal point of social and national revolutions, wars, and occupation has since Lithuanian independence in 1991 relocated itself in Europe and rediscovered its multi-cultural history. The contemporary expressions of this multiculturalism provided a point of departure for the unconventional study by de Munck et al. (2009). The authors pursued the 'otherness' of Vilnius in a panoptical space using rich and painstaking deep-ethnographic descriptions. Laimonas Briedis (2008; publisher's blurb):

> has woven the letters, diaries, utterances and reflections of strangers about this gateway city at the heart of Europe into a compelling and intimate story. The author shares with the reader a deep understanding of the Lithuanian, Polish, Russian and German identities of the place, as well as its centrality in European Jewish culture.

While works like these afford empathic insights into the living culture and heritage of the city, and Briedis in particular – as one might expect from a geographer – covers ecological aspects in some depth, the firm focus on the human stories means that the multispecies dimensions of place remain largely neglected. The provisional galleries 4 and 5 are aimed at rectifying that imbalance.

The development has been led by art historian Dr Rasa Antanavičiūtė, director of the Vilnius City Museum project and one of the authors of de Munck et al. (2009). In the first of the 'Two Rooms', the geomorphology of the city's location is presented as bare terrain without plants, inhabitants, or human-made infrastructure, hidden from everyday view. The display tells the story of how the city's landscape was formed since the Devonian (380 million years ago), how and why the city grew in its specific geological structures, and how the settlement may evolve with the potential consequences of climate change.

Connecting the city with this geological framework in more detail, the 'Vilnius Residents' gallery encourages visitors to explore the multispecies dimension of the city by interrogating their personal background as they take a kind of 'citizenship test' (see Figure 10.2), following a binary decision map

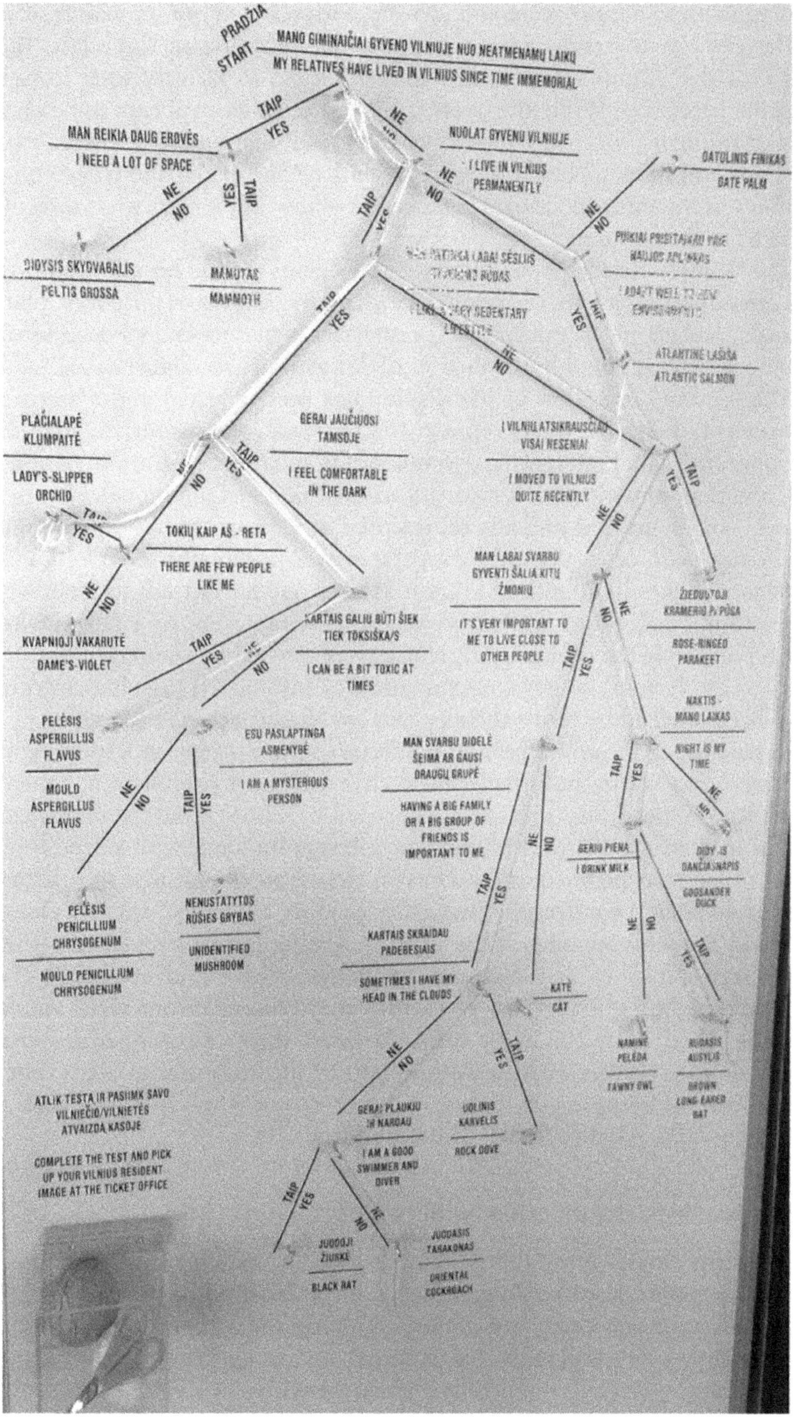

Figure 10.2 Mapping multispecies residents of Vilnius.

Source: Photograph by Ullrich Kockel (2023).

leading them to contemplate and identify with selected non-human residents of the city. At the end of their journey, they are invited to collect 'their' image of the Vilnius resident they have been led to identify with from the museum reception. While the overt focus of the gallery appears to be on the human residents, by telling their own story and relating certain aspects of it to other-than-human beings, visitors are enticed and enabled to transcend the confines of an anthropocentric perspective – how far, and in what direction, remains to be explored.

The 'Vilnius Residents' gallery takes 'residents' in the broadest multispecies sense, encompassing the populations and migrations of microbes, plants, animals, and people in a deep-time perspective including extinct species, such as the mammoth found in the district of Antakalnis (see Figure 10.3).

The human dimension of the display has both a visual and a statistical component, between them representing various other sources, including demographic data highlighting dynamics of population, migration, nationalities, and denominations. Interspersed with exhibits telling human stories are others – of plants and animals represented in various media – that connect the human with other-than-human perspectives in thought-provoking ways: physical changes in migrating salmon as they move from salt to fresh water may prompt reflections on human migrants' adaptation to a new environment; pondering the ecological significance of non-human beings regarded by humans as 'vermin' invites contemplation of instances where human others have been labelled by terms such as 'rats' or 'cockroaches'.

Reflecting on the ambition and potential of the concept underpinning this display – and taking both galleries in their intended conjunction – there is much to commend this innovative approach to representing the multispecies city in a museum context. Moving to a larger building with purpose-built gallery spaces, as planned, should enable the curators to realise their ideas in greater detail and connect the two galleries more explicitly. The one element that was arguably missing at the time of writing is part of what the Christian Creed expresses in the words 'all things, visible and invisible': These include the ghosts of the past – whether they emerged from myth and legend (see Nic Craith 2023), or from the more direct memory of traumatic interactions – who are also an integral part of the multispecies *oikos*. But on this aspect, the literature on multispecies narratives, whether in museums and more generally, remains largely silent to-date.

Cultivating New Relationships with Nature

There is an almost inevitable playfulness about contemporary attempts to develop ecological storytelling, which can be both advantageous and precarious. On the one hand, storytelling can make the complexity of ecological relationships more accessible for different age groups, levels of education, and abilities, which may contribute to increased resilience. The Vilnius City Museum project is a good example of how this can be done well, although

Figure 10.3 'Vilnius residents' gallery at the Vilnius City Museum.
Source: Photograph by Ullrich Kockel (2023).

it is too soon to assess its impact on resilience and related outcomes. On the other hand, storytelling can blur critical analysis and contribute to fantasy worlds that distract more than clarify. Critics might see the shamanic-inflected actions of Joseph Beuys, such as the controversial 'Coyote' performance, 'Explaining Pictures to a Dead Hare', or 'Celtic Kinloch Rannoch', in that category. Similarly, Kenneth White's geopoetic 'waybooks', or Ina-Maria Greverus's *flaneur* topographies, which could be regarded as newer forms of the eco-ethnographies produced by many early ethnologists, might be critiqued for not being as accessible as they could be. As examples of creative practice underpinning the postulate they share of a 'new anthropology', they constitute valuable pointers to, and elements of, the ethnological foundations for liveable ecological futures, grounded in what Abram (2020) describes as 'face-to-place storytelling', a practice that nurtures cultural renewal and replenishes a deep sense of place as a fundamental component of such liveable futures. Place is, in a sense, both the source and the destination of the ecological unity in diversity we seek, literally as well as metaphorically. It is the 'local' that Geddes, popularly alleged to have encouraged us to 'think global – act local', saw as the arena where global issues are resolved (White et al. 2023), and where critically appreciated pasts may be turned into substantive transformative futures (Siivonen and Kockel, in this volume).

In recent years, connecting heritage with sustainable development through storytelling has become one way of cultivating new relationships with nature through time. Grahn (in this volume) highlights the role of entrepreneurship in that process, anchoring the transformation in space by utilising the locality as a resource. Much of the discourse and everyday practice of sustainability remains firmly embedded in an instrumental framework of aspirations that critics (e.g., Douthwaite 1992) have long argued are ultimately unsustainable. Reframing the analysis of heritage futures from profit to gain (Kockel 2019a) can transcend the conventional binary of human-nature relationships towards a more integrated cosmology. In the process, ecological concerns and the perceptions of cultural heritage may shift and change through art (Latvala-Harvilahti and Laurén, in this volume), which can mediate contemporary encounters with historical sites (Dillon and Trenchard, in this volume) and thereby quasi change history – the stories we tell ourselves about our past – through deep engagement with place.

Mining traditional cosmologies for visions of an alternative natureculture future, Nic Craith (in this volume) emphasises the significance of language for shaping what and how we know about the places we co-relate with. Rather than being points or areas on reductionist maps, places tend to be complex composites, such as river catchments (Campbell 2021; Haanpää, Puolamäki, and Raike, in this volume), with which we relate holistically, even though elements of the whole may be particularly meaningful at an individual level of engagement. Seeking a deeper wisdom of place (ecosophy) in this sense is a hermeneutic pursuit involving creative conversation (Biggs, in this volume), and may benefit from geopoetic intuition (Modeen, in this volume). In response to the challenge of cultivating new human–ecological relationships, the Alliance for Intergenerational Resilience has hosted a number of Wisdom Councils (Williams et al. 2024) bringing together Indigenous youth with others no longer indigenous to place. In the short run, such initiatives, like the multispecies storytelling in the Vilnius City Museum, are vital endeavours to achieve socio-cultural and ecological reconciliation as a foundation for gaining common heritage futures. In the long run, let us hope with Abram (2020) that we can achieve a re-diversification of human culture, even in some sense a re-indigenising of that culture, within a sense of our larger ecological identity as Earth.

Acknowledgements

An earlier version of this chapter was delivered as a public lecture at the Vilnius City Museum on 1 October 2023. I am grateful to my online students from the 'pandemic cohort' on Vytautas Magnus University's MA in Social Anthropology who came along to see me 'in the flesh' and asked many inspiring questions that shaped the present reflections.

References

Abram, David with Tema Milstein and José Castro-Sotomayor. 2020. "Interbreathing Ecocultural Identities in the Humilocene." In *Routledge Handbook of Ecocultural Identity*, edited by Tema Milstein and José Castro-Sotomayor, 5–25. London: Routledge. https://doi.org/10.4324/9781351068840-1

Augé, Marc. 1995. *Non-Places: Introduction to an Anthropology of Supermodernity*. London: Verso.

Bahun, Sanja and Bojana Petrić. 2018. *Thinking Home: Interdisciplinary Dialogues*. London: Bloomsbury.

Basso, Keith. 1996. *Wisdom Sits in Places: Landscape and Language among the Western Apache*. Albuquerque, NM: University of New Mexico Press.

Bloch, Ernst. 1974. *Das Prinzip Hoffnung*. 3 volumes. Frankfurt/M.: Suhrkamp.

Briedis, Laimonas. 2008. *Vilnius: City of Strangers*. Vilnius: Baltos Lankos.

Bronner, Simon. 2023. "Ethnology and Folklore." *International Encyclopedia of the Social Sciences*. https://www.encyclopedia.com/social-sciences/applied-and-social-sciences-magazines/ethnology-and-folklore

Campbell, Liam. 2021. *Room for the River: The Foyle River Catchment Landscape: Connecting People, Place and Nature*. Derry: Merdog.

Casey, Edward. 1996. *The Fate of Place: A Philosophical History*. Berkeley: University of California Press.

Clingermann, Forrest. 2011. "Environmental Amnesia or the Memory of Place? The Need for Local Ethics of Memory in a Philosophical Theology of Place." In *Religion and Ecology in the Public Sphere*, edited by Celia Deane-Drummond and Heinrich Bedford-Strohm, 141–159. London: Clark.

Dakhli, Leyla, Pascale Laborier and Frank Wolff, ed. 2024, *Academics in a Century of Displacement: The Global History and Politics of Protecting Endangered Scholars*. Wiesbaden: Springer.

de Munck, Victor. 2004. "The Need for Culture (Even if it Doesn't Exist): A Lithuanian Example." *Lietuvos Etnologija: socialinės antropologijos ir etnologijos studijos* 4 (13): 33–47.

de Munck, Victor, Trini de Munck, Rasa Antanavičiūtė and Linas Svolkinas. 2009. *Experiencing Vilnius: Insider and Outsider Perspectives*. Vilnius: Kronta.

Douthwaite, Richard. 1992. *The Growth Illusion. How Economic Growth has Enriched the Few, Impoverished the Many and Endangered the Planet.* Dartington: Green Books.

Ducros, Hélène. 2021. An introduction to *Rethinking the Human in a Multispecies World*, special feature of *EuropeNow.* https://www.europenowjournal.org/2021/11/07/rethinking-the-human-in-a-multispecies-world/

Fenske, Michaela. 2023. "Narrating the Multispecies World." Opening Address at the international conference on *Narrating the Multispecies World: Stories in Times of Crises, Loss, Hope.* Würzburg, 3–5 August.

Fenske, Michaela and Sophie Elpers. 2019. "Multispecies Worlds in the Museum." *Ethnologia Europaea* 49 (2): 8–14.

Gibson, Katherine, Deborah Bird Rose and Ruth Fincher, ed. 2015. *Manifesto for Living in the Anthropocene.* Brooklyn, NY: Punctum.

Greverus, Ina-Maria. 1972. *Der territoriale Mensch: Ein literaturanthropologischer Versuch zum Heimatphänomen.* Königstein/Ts.: Athenäum.

Haraway, Donna. 2003. *The Companion Species Manifesto: Dogs, People, and Significant Otherness.* Chicago, IL: Prickly Paradigm.

Hohti, Riikka and Tuure Tammi. 2023. "Composting Storytelling: An Approach for Critical (Multispecies) Ethnography." *Qualitative Inquiry* 30 (7): 595–606.

Instone, Lesley. 2015. "Walking as Respectful Wayfinding in an Uncertain Age." In *Manifesto for Living in the Anthropocene*, edited by Katherine Gibson, Deborah Bird Rose and Ruth Fincher, 133–138. Brooklyn, NY: Punctum.

Kirksey, Eben and Stefan Helmreich. 2020. "The Emergence of Multispecies Ethnography." *Cultural Anthropology* 25 (4): 545–576.

Kirshenblatt-Gimblett, Barbara. 2019. *Agents of transformation: The role of museums in a changing world.* Keynote, 14th SIEF-Congress, Santiago de Compostela, 17 April. Abstract. Accessed 3 November 2023. https://www.siefhome.org/congresses/sief2019/keynotes.shtml

Kockel, Ullrich. 2009. "Wozu eine Europäische Ethnologie – und welche?" *Österreichische Zeitschrift für Volkskunde* 63 (4): 451–468.

Kockel, Ullrich. 2010. "Fieldpaths towards an Eco-Anthropology." *Anthropological Journal of European Cultures* 19 (2): 1–6.

Kockel, Ullrich. 2011. "Morphogenetic Fieldwork and the Ethnologic of Toposophy: Meditation on a Coyote Wandering on Rannoch Moor." In *Beuysian Legacies in Ireland and Beyond: Art, Culture and Politics*, edited by Mia Lerm Hayes and Victoria Walters, 195–219. Münster: LIT,

Kockel, Ullrich. 2012a. "Being From and Coming To: Outline of an Ethno-Ecological Framework." In *Radical Human Ecology: Intercultural and Indigenous Approaches*, edited by Lewis Williams, Rose Roberts and Alastair McIntosh, 57–71. London: Routledge.

Kockel, Ullrich. 2012b. "Toward an Ethnoecology of Place and Displacement." In *A Companion to the Anthropology of Europe*, edited by Ullrich Kockel, Mairéad Nic Craith and Jonas Frykman, 251–271. Oxford: Blackwell.

Kockel, Ullrich. 2015a. "Re-placing Europe: An Ethnological Perspective on Frontiers and Migrants." In *Museums, Place and Identity in Europe*, edited by Christopher Whitehead, Katherine Lloyd, Susannah Eckersley and Rhiannon Mason, 81–100. Aldershot: Ashgate.

Kockel, Ullrich. 2015b. "Die DEUTSCHE JUGEND DES OSTENS und die Burg Ludwigstein (1951–1975)." In *Ludwigstein – Annäherungen an die Geschichte der Burg*, edited by Eckart Conze and Susanne Rappe-Weber, 313–333. Göttingen: Vandenhoeck & Ruprecht.

Kockel, Ullrich. 2019a. "Heritage Futures: Tradition, Gain, Sustainability." Keynote address, conference on *Constructing Social Futures – Sustainability, Responsibility and Power*, University of Turku, Finland, 12–13 June. Accessed 12 September 2024. https://futuresconference2019.wordpress.com/wp-content/uploads/2019/08/ullrich_kockel_heritage-futures-public.pdf

Kockel, Ullrich. 2019b. "Commemorating Vanished 'Homelands': Displaced Germans and Their Heimat Europa." In *Heritage Festivals in Europe*, edited by Ullrich Kockel, Cristina Clopot, Baiba Tjarve and Mairéad Nic Craith, 188–204. London: Routledge.

Kockel, Ullrich. 2023. "Kenneth White: 'A Shaman Dancing on a Glacier'." In *Reading Matters*, edited by Ulrich Marzolph, 205–209. Göttingen: Universitätsverlag Göttingen.

Kockel, Ullrich. 2024. "'Are You (Or Could You Be) Indigenous?' A Perspective from Europe." In *Routledge International Handbook of Heritage and Politics*, edited by Gönül Bozoğlu, Gary Campbell, Laurajane Smith and Christopher Whitehead, 173–178. London: Routledge.

Kockel, Ullrich and Mairéad Nic Craith. 2015. "Hybride Ethnologien des Eigenen, Anderen und Dritten: Toposophische Erkundungen am Beispiel der Pfälzer in Irland." In *Das Reine und das Vermischte – 15 Jahre danach*, edited by Ines Keller and Fabian Jacobs, 73–95. Münster: Waxmann.

Kockel, Ullrich and Mairi McFadyen. 2019. "On the Carrying Stream into the European Mountain: Roots and Routes of Creative (Scottish) Ethnology." *ANUAC* 8 (2): 189–211.

Kuper, Adam. 1999. *Culture: The Anthropologist's Account.* Cambridge, MA: Harvard University Press.

Macdonald, Murdo. 2020. *Patrick Geddes's Intellectual Origins.* Edinburg: Edinburgh University Press.

MacDonnchaidh, Ruairidh and Derek Robertson. 2022. *Bho Bheul an Eòin / From the Bird's Mouth.* Balmerino: Woodlands Studios.

Malpas, Jeff and Kenneth White. 2021. *The Fundamental Field: Thought, Poetics, World.* Edinburgh: Edinburgh University Press.

Milstein, Tema and José Castro-Sotomayor. 2020. "Ecocultural Identity: An Introduction." In *Routledge Handbook of Ecocultural Identity*, edited by Tema Milstein and José Castro-Sotomayor, xvii–xxiii. London: Routledge.

Molthan-Hill, Petra, Heather Luna, Tony Wall, Helen Puntha and Denise Baden, ed. 2020. *Storytelling for Sustainability in Higher Education: An Educator's Handbook.* London: Routledge.

Nic Craith, Mairéad. 2020. *The Vanishing World of the Islandman.* Cham: Palgrave.

Nic Craith, Mairéad. 2023. "Gaelic Folklore for a Multi Species Future." *Bella Caledonia* 23 August. https://bellacaledonia.org.uk/2023/08/23/gaelic-folklore-for-a-multi-species-future/

Nic Craith, Mairéad and Ullrich Kockel. 2016. "(Re-)Building Heritage: Integrating Tangible and Intangible." In *A Companion to Heritage Studies*, edited by William Logan, Mairéad Nic Craith and Ullrich Kockel, 426–442. Oxford: Wiley-Blackwell.

O'Donohue, John. 2000. *Conamara Blues.* London: Doubleday.

Riehl, Wilhelm. 1857. *Die Pfälzer. Ein rheinisches Volksbild.* Stuttgart: Cotta.

Rose, Deborah Bird and Thom van Dooren. 2012. "Storied-Places in a Multispecies City." *HumAnimalia: A Journal of Human/animal Interface Studies* 3 (2): 1–27.

Rose, Deborah Bird, Thom van Dooren, Matthew Chrulew, Stuart Cooke, Matthew Kearnes and Emily O'Gorman. 2012. "Thinking Through the Environment, Unsettling the Humanities." *Environmental Humanities* 1 (1): 1–5.

Schulze Wessel, Martin, Irene Götz, Ekaterina Makhotina, ed. 2010. *Vilnius: Geschichte und Gedächtnis einer Stadt zwischen den Kulturen.* FrankfurtM.: Campus.

Swanson, Heather, Nils Bubandt and Anna Tsing. 2015. "Less Than One But More Than Many: Anthropocene as Science Fiction and Scholarship-in-the-Making." *Environment and Society: Advances in Research* 6: 149–166.

Tuan, Yi-Fu. 1974. *Topophilia: A study of Environmental Perception, Attitudes, and Values.* Englewood Cliffs, NJ: Prentice-Hall.

van Dooren, Thom, Eben Kirksey and Ursula Münster. 2016. "Multispecies Studies: Cultivating Arts of Attentiveness." *Environmental Humanities* 8 (1): 1–23.

Walters, Victoria. 2012. *Joseph Beuys and the Celtic Wor(l)d: A Language of Healing.* Münster: LIT.

Weiner, James. 2001. *Tree Leaf Talk: A Heideggerian Anthropology.* Oxford: Berg.

Welz, Gisela. 2021. "More-than-human Futures: Towards a Relational Anthropology in/of the Anthropocene." *Hamburger Journal für Kulturanthropologie* 13: 36–46.

White, Rehema, Ullrich Kockel, Betsy King, Kirsten Leask, Peter Higgins and Andrew Samuel. 2023. "The Future Starts in the Past: Embedding Learning for Sustainability through Culture and Community in Scotland." *Frontiers of Sustainability* 4. https://doi.org/10.3389/frsus.2023.1128620

Williams, Lewis, Rose Roberts and Alastair McIntosh, ed. 2012. *Radical Human Ecology: Intercultural and Indigenous Approaches.* London: Routledge.

Williams, Lewis, Danielle Alphonse, Ullrich Kockel and Kikila Perrin. 2024. "Transforming Approaches to Reconciliation in Higher Education and Leadership through an Inclusive Practice of Indigenist Social-Ecological Resilience." In *The Bloomsbury Handbook of Ethics of Care in Transformative Leadership in Higher Education*, edited by Mary Drinkwater, Yusef Waghid and Patrick Deane, 297–315. London: Bloomsbury.

Wolanski, Napoleon and Maciej Henneberg. 2001. "Perspectives of Human Ecology". *Human Ecology* 10: 3–7.

Index

Note: *Italic* page numbers refer to figures and page numbers followed by "n" denote endnotes.

For Product Safety Concerns and Information please contact our EU representative GPSR@taylorandfrancis.com
Taylor & Francis Verlag GmbH, Kaufingerstraße 24, 80331 München, Germany

www.ingramcontent.com/pod-product-compliance
Ingram Content Group UK Ltd.
Pitfield, Milton Keynes, MK11 3LW, UK
UKHW020140200926
475702UK00005B/833
* 9 7 8 1 0 4 1 2 1 5 8 7 5 *